# GENE AVATARS

## THE NEO-DARWINIAN THEORY OF EVOLUTION

# GENE AVATARS

## THE NEO-DARWININAN THEORY OF EVOLUTION

Pierre-Henri Gouyon
Jean-Pierre Henry
Jacques Arnould

Translated by Tiiu Ojasoo

Kluwer Academic / Plenum Publishers
New York, Boston, Dordrecht, London, Moscow

Library of Congress Cataloging-in-Publication Data

Gouyon, Pierre-Henri, 1953–
[Avatars du gène. English]
Gene avatars: neo-Darwinian theory of evolution/Pierre-Henri Gouyon, Jean-Pierre Henry, Jacques Arnould; translated by Tiiu Ojasoo.
p. cm.
Includes bibliographical references (p. ).
ISBN 0-306-46616-3
1. Evolutionary genetics. I. Henry, Jean-Pierre. II. Arnould, Jacques. III. Title.

QH390 .G6813 2002
572.8'38—dc21

2001038767

**Avatar** (Hindu Mythology): descent of deity to earth in incarnate form (*Oxford English Dictionary*). Term first used in ecology by John Damuth of the University of California.

Published in collaboration with the French Ministry of Culture—Centre national du livre.

Original edition: *Les avatars du gène: La théorie néodarwinienne de l'évolution* © Éditions Belin 1997

ISBN 0-306-46616-3


233 Spring Street, New York, New York 10013

http://www.wkap.nl/

10 9 8 7 6 5 4 3 2 1

A C.I.P. record for this book is available from the Library of Congress



Printed in the United States of America

# PREFACE

France holds a rather unusual position in the field of evolutionary biology. Whereas French naturalists from Buffon, Cuvier and Lamarck onwards made great discoveries in centuries past, French biologists missed the turning when it came to genetics. Until the 1970s, most French biologists were convinced that genetics was not as interesting as developmental science (some "rare species", for example R. Chandebois, still hold this view). For them, the general principles of heredity resided in the cytoplasm rather than in the genome. This was not, however, as great an obstacle as one might have thought. For example, it did not prevent Malécot from being one of the most outstanding theoreticians of population genetics, L'Héritier and Teissier from inventing population cages for *Drosophila,* or Monod and Jacob from discovering messenger RNA or, for that matter, from conceiving major works such as "*Chance and Necessity*" or "*Evolutionary tinkering*". In fact, it might even have been an advantage because, in France, the centralized vision of heredity rooted in Mendelian genetics was strongly tempered by the importance traditional biology conferred to the cytoplasm. Not unsurprisingly, therefore, it was in France that a mechanism whereby the cytoplasm regulates genetic information was discovered. It was in France also that the genetics of non-Mendelian elements like mitochondria and the ecological genetics of nucleo-cytoplasmic interactions (e.g. male-sterility, see Chapter 4) were developed.

Until the 1980s, evolutionary genetics was chiefly the preserve of Great Britain and the USA. These countries published all the scientific journals in the field (*Evolution, American Naturalist, Heredity, Genetics...).* Continental Europe, however, built up enough strength to found a *European Society of*

*Evolutionary Biology* and edit a society journal ("*Journal of Evolutionary Biology*") which, with time, has acquired a good reputation and substantial impact. The first editor, of American origin, worked in Basel (Switzerland); the second was the first author of this book (the editorial office is now based in the Netherlands). A major reason for creating the journal was to counterbalance the dominance of Anglo-American culture in the field. Not that our colleagues were unfriendly. On the contrary, they helped us a lot in developing evolutionary genetics in France. We are deeply indebted to scientists like Bob Allard, Janis Antonovics, Herbert Baker, Arthur Cain, Deborah Charlesworth, Theodosius Dobzhansky, Chris Gliddon, John Harper and Ledyard Stebbins - among many others - for their advice and collaboration. The trouble was that not enough account was taken of the "cultural" differences between "Anglo-Americans" and Europeans in a field of study - evolution - with such close links to ethics and politics. (We shall not discuss the ambiguity of the position of Great Britain which belongs to both worlds).

Evolutionary biology tells us that isolation, by preventing a group of individuals from being drowned in the average mass, can foster novelty. Protection from mainstream genetic and evolutionary thinking has certainly helped French biologists retain their strong interest in development. In France, it was obvious - maybe too obvious - that a synthetic theory of evolution had to take ontogeny and non-Mendelian genetics into account long before Gould and Lewontin made such a claim. The original French edition of this book was written in order to acquaint the French reader with the theory of evolution. To the English-speaking reader its interest lies not in any new theory but in the rather unfamiliar emphasis.

The authors' interest in evolutionary genetics was awakened by Georges Valdeyron's course on population genetics at the "Institut National Agronomique" in Paris. This book is dedicated to him. The French edition benefited from the ideas of Anne Atlan, Denis Couvet and Isabelle Olivieri and was improved by suggestions from Beatrice Albert, Irène Till-Bottraud, and many others. The authors are also grateful to John Maynard Smith and Andy Dobson for their judicious comments.

**Pierre-Henri Gouyon**
*Professor at Paris-Sud University*
*Lecturer at Ecole Polytechnique*
*Consulting Professor at INAPG*
*Director of UPRESA 8079*

# TABLE OF CONTENTS

# GENE AVATARS

## THE NEO-DARWINIAN THEORY OF EVOLUTION

# Introduction

## THE BIOLOGIST'S DREAM

In the 4th century Saint Augustine wrote: *"What then is time? If no one asks me, I know; but if I am asked and wish to explain, I do not know."* [1] Augustine's query about time seems to apply also to life. Can any biologist claim to know the ultimate definition of life? If he did, he might be offered an honorary chair on the National Ethics Committee but it is more likely that he would be in real danger of taking himself for yet another Faust or successor to Dr. Frankenstein. Would he not have acquired the knowledge and thus the power held by the gods? Would he not have become God?

These epistemological and metaphysical issues have not stopped biologists from pursuing their quest to understand living things, nor prevented the deep changes that have occurred during the last two centuries in a field of biology - reproduction and heredity - that has always fascinated mankind. We are far from Montaigne's outcry (Essays, Book II):

> *What kind of monster is the drop of seed from which we are produced and that encapsulates the impressions not only of bodily form but also of the thoughts and inclinations of our fathers? Where does this drop of water harbour these infinite forms? and how does it pass on those similarities that progress in so bold and disordered a fashion that the great-grandson looks like his great-grandfather, the nephew like his uncle?*

Jean Rostand (1894-1977), a French biologist and prolific author, considered this passage by Montaigne exceptional on many counts. It asks how not only physical characters are passed on but also mental traits. It conveys the process' capricious nature and the onlooker's feeling of surprise and alarm. It is a milestone we can use to judge how far our understanding has progressed over the last four centuries. Modern science has unveiled many of the secrets of reproduction and provided answers (albeit rudimentary ones) to the questions: "Why are we - living beings, human beings - here?", "Why are we born to die?", "Is there a reason to life?" It has also provided tools for modifying human inheritance, i.e. the genome. Part of the biologist's dream is embodied in the techniques of genetic engineering.

Some of our contemporaries fear this dream will turn into a nightmare. Where will biological science and techniques take us? Might they not reach down to the very foundations of life and shatter them? The spectres of Hiroshima and Nagasaki still haunt many minds, imparting virulence to the claims of ecologists and increasing fear of all that is undertaken in genetic engineering. There are Ethics Committees everywhere, in hospitals, research departments, government institutions, and international organisations. Ethical issues are being introduced into the syllabuses of schools, universities and polytechnics. Our intention is not to discredit these initiatives; all efforts to make research-workers, practitioners, and citizens more conscious of - and responsible for - their individual and collective behaviour are welcome. But the risk is that science will no longer be seen as a provider of benefits but as a sterile, expensive, and dangerous activity.

This book does not address the ethical aspects of modern genetic and biological techniques but describes the contemporary biologist's insight into life. It defends the evolutionist and especially neo-Darwinian paradigm. Charles Darwin's work was a radical turning-point in the history of biology; its quality and importance to biology cannot be called into question. In the words of John Maynard Smith, *"ever since Darwin, the theory of evolution has been the main unifying idea in biology"*.[2] We use the word paradigm in the sense defined by the physicist and historian of science, Thomas Kuhn, in *The Structure of Scientific Revolutions*.[3] A concept becomes a paradigm if the work that introduces the concept (in this case, the virtually mythical *The Origin of Species*) is so exceptional that it draws a cohesive group of researchers away from rival modes of scientific activity and provides them with all sorts of new problems to resolve.

Will the neo-Darwinian paradigm one day meet with the fate of historical monuments condemned to ruin or restored so conservatively that all later transformation is impossible? Certainly not. This paradigm embraces all past and present research activities which help *"throw some light on the origin of species - that mystery of mysteries -"* as Darwin called it.[4] The advances in genetics and ecological sciences [5] over the last few decades have already had a strong influence on the evolutionist viewpoint and brought about changes even though many questions still need to be answered. It is intellectually dishonest, therefore, to mistake current scientific controversy and debate for a radical attack of the neo-Darwinian paradigm and to propose, even impose, a "new paradigm" which is most often unscientific and whose problem posing and solving modes are hardly any different.

Within the neo-Darwinian paradigm, the issue of what is life has a tone and expression of its own. On what entities, how, and at what integration levels does natural selection act? Why do we speak of the "selfish gene" and what does this mean? What is a living being? These questions are answered in Chapter 4 but, to understand the answers, we have to put them into their historical and epistemological context (see Chapters 1 to 3). The historical roots of current biological concepts run very deep; the connotations and implications of these concepts are often variable, as we shall illustrate when discussing what is a gene. To forget this is to leave the door open to misunderstanding, ignorance, and even injustice as these scientific discoveries have weighty social implications. The first applications of genetics were endeavours to improve the human species. Eugenics is, so to speak, the "atomic mushroom" of modern biology.

Once the concepts are explained and put into context, we shall find that they raise questions which prompt a reappraisal of living things. What are the relationships among elements of the genome, members of a population or society, and species of a community? What is the reason for sexual reproduction, for death? Chapters 5 and 6 will tell us how the neo-Darwinian paradigm tackles these issues by drawing upon evolutionary genetics.

The last chapter addresses the aspects of current evolutionary biology we consider most fundamental. There is no doubt that contingency - history's imprint on the living world (what happens does not depend upon a necessity that transcends history but upon all that precedes) - and constraints, whether anatomical, physiological or other, are involved in the processes at the origin of living beings and mankind. Many controversies oppose evolutionary biologists in this area simply because there are major obstacles to our further understanding of the living world. These obstacles prevent the ultimate dream of the biologist - foretelling the past - from coming true.[6]

**NOTES**

1. Augustine, Confessions, Book XI, Chapter 14. (*"Quid est ergo tempus? Si nemo ex me quaerat, scio; si quarenti explicare velim, nescio"*). André Pichot makes the analogy to life in *Histoire de la Notion de Vie*, 1993, Gallimard, Paris.
2. Maynard-Smith, J., 1989, *Evolutionary Genetics*, Oxford University Press, Oxford, p. V.
3. Kuhn, T.S., 1970 (first published in 1962), *The Structure of Scientific Revolutions* (2[nd] edition, enlarged), The University of Chicago Press, Chicago, p. 10.
4. Darwin, C., 1859, *On the Origin of Species by Means of Natural Selection or the Preservation of Favoured Races in the Struggle for Life*, John Murray, London.
5. In 1886, Ernest Haeckel called the science of the relations between living beings and the world that surrounds them "ecology". Ecology is thus not confined to just the "functional" dimension of biological processes and of the relationships these processes establish

among living or inert elements of the biosphere. It also takes the historical dimension into account, especially of the evolution of living beings and of the groups (populations, species) they form.

6. In other words, providing a full explanation of how one biological form in the past gave way to another, before this passage is illustrated by the discovery of fossils.

# Chapter 1

## "WAITING FOR DARWIN"
## From pre-Socratic myths to the transformism of Lamarck

Just by looking at plants or animals grow, we are prompted into asking intriguing questions about the fate and history of living things. These questions arise from the prodigious unity and harmony of biological processes and also from the human mind's difficulty, even inability, to grasp all the underlying mechanisms. They voice at least three concerns. The first is about the origin of self and mankind (a wise man in the Bible says: "*There are three things which I cannot understand, and a fourth which I know not - the tracks of an eagle in its flying - the footsteps of a serpent on a rock - the marks of a ship crossing the sea - and the ways of a man in youth*" [1]). The second is about daily rations (in Antiquity, it was widely believed that, on poor soil, wheat could change into barley just as barley could change into oats), and the third about the future.

**Philosophical and religious views on the origin of the world**

When considering the origin of the world from a philosophical and religious standpoint, we have to distinguish between procession, emanation, transformation, and creation. Procession takes place when an immutable character is transferred in its entirety to several beings without any substance being divided. Emanation occurs when a being derives from its own substance a similar or analogous substance with a separate existence, or when a being produces within itself a new manner of being that is both distinct (independent) and indistinct (self-supported). Transformation occurs when an external agent intervenes and changes the state of a being. Finally, there is creation when God confers existence, without himself, to a substantiality that did not pre-exist.

All cultures have their own folklore about the origins and metamorphoses of living things as this tale from a 17th century French botany book:

*There is a tree admittedly not common in France but often found in Scotland. Leaves fall from this tree. On one side, they strike water and turn into fish; on the other, they strike the ground and turn into birds.*[2]

Just like myths, these imaginary tales try to explain the origin and meaning of the world and how it came to look the way it does. They have given way to philosophical and scientific theories in all the cultures where these theories have developed and taken a hold. However, representations of the world in both mythological or scientific discourse have always embraced the idea of evolution even though this idea may have taken very different forms and held different positions in time and place.

*Figure 1.1* Picture of the tree whose leaves, once they have struck the ground, turn into flying birds. The leaves which hit the water turn into fish. Reproduced from Duret, C., 1605, Histoire Admirable des Plantes.

# 1. BEFORE EVOLUTIONISM: METAMORPHOSIS, FIXISM, SPONTANEOUS GENERATION

## Evolutionism before Socrates

The cosmogonies of Chinese and Indian culture are without doubt "evolutionist".[3] They speak of a world that arose from a single, simple source, and then differentiated. In general, most myths involving primitive elements have a common scheme. Starting from earth, water, air, fire, or some other initial state, Nature gradually acquires new properties, sometimes at the expense of earlier equilibria. However, Nature is only outwardly or partially renewed since her initial unity remains implicit and may even become more apparent.

An analogous view was held by pre-Socratic thinkers and even predominated in their, alas fragmentary, works. These gradually set aside most mythical elements. Nature was seen as a living being, an organism driven by a life force, a force that leads to generation and development and that transforms the organism itself. For example, Anaximandrus (610-547 BC) and Anaximenus (ca 550-480 BC) claimed that living beings were born through spontaneous generation by the action of the sun on a humid starting material. According to Anaximandrus,

> *The first animals were born through humidity and were encased by a thorny skin. As time went by, they crept up onto the bank; their casing unlocked and shed, they quickly went on to live a new way of life.*

Generation is obviously spontaneous when we watch with a child's eyes the emergence of flies and maggots. Anaximandrus saw spontaneous generation not only in the simplest beings but also in plants and higher animals. He allied to it the idea of metamorphosis as when an insect passes from the larval stage to adulthood (thereby drawing an interesting parallel between phylogeny and ontogeny, between the beginnings of mankind and the beginnings of an individual).

> *Firstly, man was born from living beings of a different kind because, unlike animals who start to hunt early, he needs prolonged care during early childhood. If he had started off in this state, he would not have survived. Thus, man first took shape in creatures that resembled fish and stayed there like an embryo until maturity. The creatures then burst open, and men and women able to care for themselves came forth.*[4]

Once spontaneous generation and metamorphosis had been brought together, it became possible to imagine how human forms had evolved over time and to explain how extravagant freak forms had emerged in Nature. Most pre-Socratic thinking on life rested on the notions of randomness, chance, and accidental encounters, as well as on the notions of need - even foresight -, planning, and goal. This is not unlike the "modern" evolutionist concept of the living world which sees finality no longer as a purpose at work within evolution but as an unavoidable feature of the structures created by natural selection (see J. Monod's view of "teleonomy" in Chapter 7).

Empedocles also espoused the idea of spontaneous generation. At the beginning, separate pieces emerged directly from the wet earth and clay: single limbs, heads without horns, faces without foreheads, arms without shoulders, and eyes without a face. These pieces then assembled randomly, in all possible combinations, to form all kinds of monsters: bulls with human heads, men with bulls' heads, two-faced animals, creatures with countless hands, and so on. Some pieces remained unattached. This free assembly phase was the era of friendship and was followed by the era of hate and sexuality. Disharmonious fauna disappeared, and animals were no longer born directly from the earth but were generated. Thus, Empedocles envisaged a succession of fauna in which defective combinations were eliminated; the hallmark of life was true unity and affinity among beings.

For Democritus (ca 460-370 BC), the founder of atomist theory, Nature was formed by the haphazard motion of atoms in a vacuum. The human being is the fruit of chance (there is no finality). He creeps out of the water and silt "*like a small worm*". Democritus extended his view of Nature to society and culture. This was an echo of the golden age of many traditions or, on the contrary, of the time when mankind mastered no technique nor art and led a precarious life. He thought that language might have evolved from sounds that were no more elaborate than the cries of animals, but that were more effective.

The idea of a creative chance, also present in Epicurus' philosophy (341-270 BC), was taken up by Lucretius (95-53 BC). He too saw the earth as the begetter of all living things. "*It was not by a golden cable*, he wrote, *that living beings descended down to earth from the sky*." The earth first gave birth to grasses and small trees, then to many animal species. Of the monsters that thus came to be, only those that could move about, feed themselves, and reproduce, survived. Finality in the organism is accidental; in other words, "*we do not have eyes to see, we see because we have eyes*". Lucretius was the first to introduce the notion of a struggle for life among living beings.

> *All those whom you see breathe the air that sustains them have relied on either stealth, vigilance, or agility to defend and protect their species right from its beginnings... Lions prevail through courage, foxes by subterfuge, deer by flight..... Domestic animals were protected by man....*

*But the creatures that nature has not endowed with self-sufficiency or qualities of use to us, or any reason for being fed and protected by us, were exposed as food for others, encumbered by the fettles that fate had forged for them, and perished to the last by Nature's command.*

All these thinkers not only recognised that chance and monstrosities exist in the living world but also endowed chance with a kind of procreative power. This is contrary to the thinking of Plato and Aristotle for whom monstrosities were just exceptions to a world order.

## Greek essentialism. Creationism in the Christian tradition

Plato (428-348 BC) and Aristotle (384-322 BC) inaugurated the era of essentialism. In the *Timea*, one of Plato's richest dialogues, the demigod created the Universe with the help of the model provided by the world of Ideas. All forms of life thus created are imperfect copies of these types or essences which alone are genuine, unchanging, and perfect. All of Nature is taken to be a living being [5] composed of other, different kinds of living beings. First, there are the gods, created directly by the demigod, then aerial, aquatic and terrestrial animals. The human male is the most accomplished terrestrial being whereas other animals are just degraded forms (including woman!). Plato does not mention plants.

**The work of the demigod according to Plato**

*Thus far and until the birth of time the created universe was made in the likeness of the original, but inasmuch as all animals were not yet comprehended therein, it was still unlike. What remained, the creator then proceeded to fashion after the nature of the pattern. Now as in the ideal animal the mind perceives ideas or species of a certain nature and number, he thought that this created animal ought to have species of a like nature and number. There are four such; one of them is the heavenly race of the gods; another, the race of birds whose way is in the air; the third, the watery species; and the fourth, the pedestrian and land creatures. Of the heavenly and divine, he created the greater part out of fire, that they might be the brightest of all things and fairest to behold, and he fashioned them after the likeness of the universe in the figure of a circle [...]*

*All these the creator first set in order, and out of them he constructed the universe, which was a single animal comprehending in itself all other animals, mortal and immortal. Now of the divine, he himself was the creator, but the creation of the mortal he committed to his offspring.*[6]

Plato's view of the world is said to be essentialist because it is founded on an array of types or essences. It can be the basis of a taxonomy; types can be classified because they are quite distinct and do not change over time. We are reminded of the current method of authenticating a new species by depositing its type in a museum.

Aristotle was more interested in the hierarchy of living beings. He did not speak of "Life" as such, but of the "Living". He admitted that his typology was imperfect and not easy to use because many intermediate forms smudged the boundaries. However, his universe was also eternal and fixed, as was the nature of each of its components. He rejected the idea of passage from one species to another or of the emergence of complex forms from simpler ones. Motion was an essential feature of living things but was only an expression of their power and could not induce any real change in cosmic order.

**The fixity of the world according to Aristotle**

*... for any living thing that has reached its normal development and which is unmutilated, and whose mode of generation is not spontaneous, the most natural act is the production of another like itself, an animal producing an animal, a plant a plant, in order that, as far as its nature allows, it may partake in the eternal and divine. That is the goal towards which all things strive, that for the sake of which they do whatsoever their nature renders possible. The phrase 'for the sake of which' is ambiguous; it may mean either (a) the end to achieve which, or (b) the being in whose interest, the act is done. Since then no living thing is able to partake in what is eternal and divine by uninterrupted continuance (for nothing perishable can for ever remain one and the same), it tries to achieve that end in the only way possible to it, and success is possible in varying degrees; so it remains not indeed as the self-same individual but continues its existence in something like itself - not numerically but specifically one.*[7]

Both Plato and Aristotle, despite their diverging views on the origin of the world, defended its unchanging character and rejected any idea of evolutionism. Their legacy - that the world has a hierarchy that is fixed - can be reconciled fairly readily with the faith in a "God-the-Creator" professed by followers of the Jewish and Christian religions in the Western world. Both these religions consider that, in the very beginning, God created a complete world from nothing (*ex nihilo*). Creationists claim that all the types within the created world existed, either explicitly or potentially, right from the beginning. Christian thinking, even though it defends the idea that there is a history and that time flows irreversibly, is essentialist. Mankind has the key role of co-creator responsible for maintaining order and for ensuring that history unfolds satisfactorily within creation. It is not hard to see why, within such a tradition, all attempts to consider evolution were doomed to failure.

**What is meant by "creationism"?**

Creationism can refer to the Christian tradition; we speak of God the creator. It can also refer to the Jewish and Islamic traditions. Since the mid-19th century, however, its use has been restricted to the fundamentalist movements which contest the theories of Darwin and his followers and which defend the idea that the world is not the result of the slow workings of natural forces such as selection, adaptation, or chance, but the work entirely of God, as recounted by Moses, through divine revelation, in the first chapters of the book of Genesis.[8]

Religion does not, however, reject all kinds of variability. In his commentary to the first chapter of the Book of Genesis, Augustine reflects on how different life forms appeared in turn. His thinking is based on the Stoics' doctrine of seminal reasons. Vital energy, albeit one and infinite, is present in an infinite number of confined bodies as seeds that slowly give shape to matter. The origin of life forms can thus be explained by a generation model. An entirely formed living being pre-exists in the seed. This view prefigures pre-formationist ideas (see below). Through Plotinus, Augustine introduces some of its aspects into the Christian tradition and considers the history of creation as a gradual unfolding of a divine plan.

In the 13th century, Thomas Aquinas saw the world of living organisms as a ladder reaching from the simplest to the most elaborate forms of life. He borrowed the idea from Aristotle's Ladder of Nature. In fact, during early Christianity and in the Middle Ages, Western thinkers probably believed less in biological fixity than did their predecessors. They were convinced that man could be changed into a were-wolf just as vile metal could be converted into gold. All these opinions hinged around a belief in spontaneous generation.

## Was the fixity of species the price to pay to do away with spontaneous generation?

The idea that some living beings might arise spontaneously actually derives from errors in observation. The (presumed) origin of barnacle geese is a picturesque example of such an error. Barnacle geese are maritime geese which migrate by night and show up suddenly on the coast of Western Europe. To the people of the Middle Ages, their sudden arrival was explained by the fact that they were born from flotsam. They came upon the idea by observing the barnacle, a cirripedia crustacean. The planktonic larva, *nauplius*, clings to floating wood at sea and undergoes metamorphosis into a pedunculated organism. Its main body valves let protrude feathery brachia and its foot looks like the flexible neck of a goose. Popular science needed no other proof for the spontaneous generation of barnacle geese!

The idea that complex organisms such as vertebrates might be generated spontaneously lost much support with time; this was not the case for lower life forms. Until the 17th century it was commonly believed that flies were generated by urine or by rotting meat. In 1668, Francesco Redi, poet, antiquary, physician and naturalist, performed the first experiments that contradicted this belief. Redi placed samples of snakes, fish, and veal in closed and open jars. He chose veal probably not because he had a liking for this meat but because, according to popular belief, veal was especially prone to engendering flies. However, it was not Redi's choice of material that makes his work noteworthy but his excellent idea of including experimental controls. He wrote to his fellow-members of the *Accamedia del Cimento*:

> *All our efforts must focus on the experiments, on the creation of standards of measurement and on the elaboration of exact research methods.*[9]

The result was unmistakable. The samples in the open jars, but not those in the closed jars, were ridden with maggots. Redi concluded:

> *The flesh of dead animals cannot engender worms unless the eggs of the living be deposited therein.*[10]

Notwithstanding, it took nearly two centuries for the theory of spontaneous generation to be definitively rejected as an explanation for living organisms.

In 1748, John Needham repeated Redi's experiment with more care and precision. His results were completely at variance with Redi's as were his intentions. Needham prepared a meat broth and checked under the microscope that it contained no organisms before placing it in a sealed jar. When the jar was opened a few days later, it swarmed with thousands of organisms. Was this, at last, proof of spontaneous generation?

The abbot Spallanzani (1729-1799) replicated the experiment arguing that the creatures observed were so small that their eggs (or spores) must have been smaller still and must have escaped Needham's attention. Spallanzani first demonstrated that the "little animals" could appear in an air-tight vial containing relatively little air. He then filled vases with various infusions, sealed them over a flame, and plunged them in boiling water for an hour. No "little animal" appeared, at least whilst the vials stayed closed.

The results of his first experiment excluded the action of heat as a rarefier of air but did the outcome of his second experiment really demonstrate that the proliferation Needham observed was due to pre-existing germs? Some onlookers claimed that the air, once heated, could not support the life that was supposed to have appeared spontaneously.

It was still a topical debate in the middle of the 19th century when the Paris Academy of Sciences decided to offer a prize to the person who would settle it once and for all. In 1859, a French scientist, Félix Pouchet, published a 700-page opus in which he set out to demonstrate that the theory of spontaneous generation was sound. He not only reviewed the most advanced ideas but also substantiated his overview with many experimental facts. His demonstration was short-lived. The controversy was settled in 1862 when Louis Pasteur provided undisputed proof for large quantities of microscopic germs in the ambient air and also on the hands, apparatus and utensils of those who perform experiments. There could be no question of spontaneous generation; it was a matter of contamination of culture broths. Microorganisms are the agents of putrefaction and not the products of the death and decay of organic matter.

Why have we discussed the belief in spontaneous generation at such length? Simply because it was instrumental in delaying acceptance of the idea of evolution in biology. As long as common belief accepted the likelihood of spontaneous generation and metamorphosis, there was no room for evolution, for the idea that different organic forms can be linked by means of generation. To dismiss belief in spontaneous generation and species interconversion and claim that species are fixed would have meant having to address the issue of species constancy over historical time and over generations. When we look back, we realise that this fixist phase was, of course, necessary in order to broach the idea of the evolution of life later on. Redi bears testimony to this turning point in the history of ideas on life when he writes:

> *I shall express my belief that the Earth, after having brought forth the first plants and the first animals by order of the Supreme and Omnipotent Creator, has never since produced any kind of plants or animals, either perfect or imperfect; and everything which it has been known to produce in past or present times came solely from [...] the seeds of the plants or animals themselves; by their own means, they thus preserve their species.*[11]

## And yet, it does go round...

There is a touch of irony to the echoes that resound through history.

1633: Galileo had to retract before the tribunal of the Inquisition. Legend has it that he whispered to himself: "*And yet, it does go round!*".

1668: Redi provided the first proof against spontaneous generation and in favour of the fixity of species.

1742: Linnaeus asked himself: "*And if, after all, they [species] change?*"

Carl Linné, alias Linnaeus (1707-1778), championed the idea of the fixity of species. His outlook on species was essentialist and close to a Platonist's view of the world. If there is an ideal type for each species, then deformations, imperfections, monstrosities are just variations of no importance. Linnaeus' view of what we now call biodiversity was thus based exclusively on variations among species with no concern for within-species variation. (It is a view that some of our contemporaries still hold on the wealth and variety of the living world.)

**The fixity of species according to Linnaeus in 1737**

*All species reckon the origin of their stock in the first instance from the veritable hand of the Almighty Creator: for the author of nature, when He created species, imposed on his Creations an eternal law of reproduction and multiplication within the limits of their proper kinds. He did indeed in many instances allow them the power of sporting in their outward appearance, but never that of passing from one species to another. Hence to-day there are two kinds of difference between plants: one a true difference, the diversity produced by the all-wise hand of the Almighty, but the other, variation in the outside shell, the Work of nature in a sportive mood. Let a garden be sown with a thousand different seeds, let to these be given the incessant care of the Gardener in producing abnormal forms, and in a few years it will contain six thousand varieties, which the common herd of Botanists calls species. And so I distinguish the species of the almighty Creator which are true from the abnormal varieties of the gardener: the former I reckon of the highest importance because of their author, the latter I reject because of their authors. The former persist and have persisted from the beginning of the world, the latter, being monstrosities, can boast of but a brief life.*[12]

Linnaeus thus dismissed all concessions to variability involving spontaneous generation. Species were not the product of a blind process prone to accidents but they reproduced a type fixed by "God". By defending the fixity of species, Linnaeus was able to elaborate a genuine system of classification, the one that made him famous. In 1735 he published the first edition of *Systema naturae, sive regna tria naturae, systematice proposita per classes, ordines, genera et species*, a 14-page in-folio that was to have 14 editions, the last comprising 10 volumes. This work heralded the reign of natural history, and the era of world expeditions to find and collect flowers, birds, animals, butterflies and shells. No species was to escape the notice of the flourishing trade in specimens! The intense activity in taxonomy and the vogue for natural history study-rooms were the consecration of Linnaeus' doctrine.

And yet, might species not undergo change? In the 18th century, the first serious doubts about fixism were raised even though it had helped much in furthering understanding of the living world. Linnaeus might well make a distinction between diversity of divine origin and diversity due to a gardener's whims, there was nevertheless the troublesome point that horticulturists and breeders could derive new forms.

In 1742, a student brought Linnaeus a strange, monstrous plant which he baptised *Peloria* (from the Greek *pelorios*, prodigious, monstrous). Its general habit, stem, leaves, roots, and fragrance, as well as the colour of its flowers, were similar to those of common toadflax (*Linaria vulgaris Miller*) but the flowers were of a completely different shape. *Linaria* had just a single spur at the base of the corolla; *Peloria* had five. Why? Linnaeus suggested that new species were generated by hybridisation, and one of his students summed up his hopes as follows:

> *It would be the delight of our century to have discovered phenomena that were not only unknown to our forebears but that are hardly to be believed.*[13]

In 1762, Linnaeus, still rather puzzled, suggested that God had created genera and that species had appeared afterwards. Even later on, he accredited the idea that all species of the same genus were originally varieties of a main species which had multiplied through hybrid generations.

> *For a long time I have suspected, but I dare not put forward the hypothesis, that all species belonging to the same genus originally formed a single species which diversified by hybridisation. There is no doubt that this is one of the issues of the future and that many experiments will be set up to convert this hypothesis into an axiom establishing that species are the work of time.*[14]

The *Peloria* was not an isolated example. In 1715, about 50 years before Linnaeus' time, Jean Marchant, head of the nurseries of the King's Gardens (*Jardins du Roy*), discovered two "species" of *Mercurialis annua L.* These had leaves that were more or less lacinate and that were very different from those of the common dented variety of the plant. Having succeeded in multiplying these two kinds of *Mercurialis* from seed, he was quite sure that he had witnessed the birth of new forms and, four years later, wrote to the Paris Academy of Sciences:

> *One could therefore imagine that, once the Almighty had created individual plants as models for each genus [...], these models, when they became established, produced varieties; those varieties that remained constant and permanent became species. In time and in the same fashion, these species gave rise to other products that were different.* He added: *Every day, art, culture and especially chance, that is to say unknown circumstances, generate novelties among peculiar flowers such as anemones and buttercups, and botanists consider these novelties as mere varieties that do not warrant modifying the species. But why should Nature be incapable of producing novelties that would warrant modifying the species?*[15]

His conclusion was more daring than that of Linnaeus for whom hybridisation phenomena were at the origin of new species.

In a similar vein, in 1766, the horticulturist Antoine-Nicolas Duchesne observed the appearance of a unifoliate strawberry plant which he reproduced from seed and sent to several colleagues.[16] The leaves of this plant had a single leaflet instead of the usual three, and this character was fixed and transmissible. In the words of Duchesne, the existence of varieties "*will, of course, always be evidence contradicting the opinion that species are unalterable.*" It occurred to him that the various strawberry plant varieties might be affiliated and arise from the same primitive form. He therefore decided to trace their genealogy, pointing out that:

> *The genealogical order is the only one indicated by nature, the only one that fully satisfies the intellect. All others are arbitrary and meaningless.*[17]

Was this the end of the reign of fixism and essentialism? No, not yet, even though the close of the 18th century did witness the emergence of the idea of evolution without a return to pre-Socratic myths (no doubt because of the criticisms waged against spontaneous generation). This revival of an old idea was due less to the doubts that befell Linnaeus than to the hypotheses formulated by Marchant and Duchesne.

# 2. THE CONCEPT OF EVOLUTION. A FRENCH DISCOVERY

## Modern philosophers and evolution

It would be a mistake to believe that the idea of evolution took modern thinking by storm. It was an idea that gained recognition gradually, based as it was on the observations and hypotheses of scholars and on philosophical and theological questioning.

Our modern era has seen the birth of a new cosmogony; it calls upon the laws of mechanics to uphold theories about how the universe came into existence, and has replaced cyclical time by the linear time of physics. Abiding by the Judaic and Christian traditions, Descartes, Kant and Laplace introduced the notion of history into the philosophy of nature and erased the last remaining traces of the pre-Socratic view of perpetual recurrence. In addition, two currents of thinking, the doctrines of continuity and nominalism, brought indirect support to the development of evolutionist ideas. In different ways they both ran counter to the essentialist view of life inherited from Plato and Aristotle.

Understanding life in the continuity mode means finding similarities among living creatures and building links by homology or analogy. In other words, the discontinuity that the doctrine of perfect types perceives among different forms does not exist. In *Criticism of Judgment* (1790), Kant noted that, because we can make an analogy among the forms of various organisms, we assume that there is a common type from which all these forms are descended. According to the principle of continuity which he analysed in *Criticism of Pure Reason* (1781), the necessary transitions cannot be jumps (*saltus*) but must be just very small variations. He added, however, that this continuity in forms was still only an idea that needed to be put to the test.

Nominalist thinking questioned the very existence of essences and only recognised the existence of individuals. General terms were thus only signs referring to singular entities.

Anatomists were also studying generation phenomena at this point in time. They belonged to two schools. The preformationists claimed that the female egg (or, for some, the spermatozoa) encased a miniature adult with "preformed" parts and organs (consequently, the egg also contained the embryo of the next generation, etc.)[18] On the other hand, the partisans of epigenesis believed that the embryo was made up of elements of crude matter which gradually organised themselves. The relationship between these two schools of thought and earlier views is rather complex. The preformationist hypothesis with its ready-made descendants did not accept variations due to chance. Epigenesis was a restitution of the essentialist view because an invariable specific form somehow mysteriously "told" matter how to make the embryo; variations were just in-process blunders.

## Fossils : From the *lusus naturae* to the archives of the earth

The term "fossil" first appeared in the French language in 1556. Its Latin root is the verb *fodere*, to dig, whence *fossilis*, "extracted from the earth".[19] Fossils were already known in Antiquity though there was no clear understanding about their origin.

According to Herodotes (ca 484-420 BC), petrified shells were mentioned by Egyptian priests who attributed their origin to the sea and concluded that the land of Egypt had once lain beneath the sea. The geographer Strabo (ca 58 BC - 21~25 AD) noted that such shells were sometimes found around the Mediterranean two to three thousand stadia inland, as in Armenia or Phrygia. Wasn't this proof that these areas had previously been occupied by the sea? Plutarch (ca 50-125 AD) added that, in Egypt, there were shells up mountains and down mines. Ovid (43 BC - 17/18 AD) wrote:

> *I have seen the ocean where before there was firm land; I have seen land spring from water and, far from the coast, the earth is covered with sea-shells.*[20]

On the other hand, when Xenophanes of Colophon, at the close of the 6th century BC, found the petrified remains of fish in Malta, Sicily, and Paros, he concluded that the silt that had once covered the surface of the earth had borne life. Finally, to Plinus (23-79 AD) fossils were just stones produced by lightning or curiosities of nature (in Latin, *lusus naturae*); nature takes delight in making mere pebbles look like shells, leaves or fish!

This opinion was long-lived but, at the end of the 15th century, Leonardo da Vinci (1452-1519) expressed doubts as to whether "figurative stones" were born on hill tops under the influence of stars:

*I wonder where they are now, the stars that once created shells of different ages and aspect on hillsides?*[21]

At about the same time, in 1517, Frascator claimed that fossils were the remains of animals that had once been alive. At the end of the 16th century, Bernard Palissy (ca 1510-1589/1590), already famous for having invented ceramics, set up a physics and natural history study-room in Paris. He thought that the fossil-rich layers were due to the sedimentation and petrification of fishes and shells in ancient seas but rejected outright the explanation put forward by his contemporaries - naturalists and theologians - who saw in these petrified remains the sign of the Great Flood described in the Bible. The diluvian theory had a long-lasting influence on the interpretation of fossils and the perception of the Earth's history, showing how difficult it is to imagine an ancient world different from ours. Thus, when F. Plater discovered the fossilised bones of an elephant near Lucerne in 1641, he concluded that they were the remains of the skeleton of a very tall man.

Collecting fossils was a great 18th century hobby, as witnessed by two works of rather unequal merit, *Telliamed* and *History of the Earth*. *Telliamed* is the anagram of the name of the author, Benoît de Maillet (1656-1738) who was Consul of France in Egypt, then Inspector of the commercial ports of Barbary and of the Levant. In his book, Maillet claimed that all terrestrial animals derive from fish. To him, all the land of our planet had emerged from the sea. When the sea fell back, only aquatic animals were left on firm land; only those that underwent transformation survived.

**The origin of animals in the Telliamed**

*If we now turn to the issue of the origin of animals, I notice that there is no creature that walks, flies or creeps for which there is not a similar or approaching species in the sea; its passage from the one element to the other is possible, probable, and even supported by many examples.*

*I think, Sir, the Philosopher continued, that I have given proof enough for a system where terrestrial animals derive from marine animals, where they are formed naturally in the sea from water-borne seeds whether these seeds are eternal or created (since you accept creation). It is thus easy to understand how all living, feeling and vegetative things can be generated on an Earth undergoing re-population or not yet populated.*[22]

Voltaire showed a lively interest in the works of contemporary biologists but chiefly to criticise them and make scathing remarks. He never missed an opportunity to mock de Maillet, for instance in *The Singularities of Nature* (1768), *The Systems* (1772) or the *Dialogues of Evhémère* (1777). In *The Man of Forty Ecus* (1768), he took issue with those "*who created the universe with their pen, as God had done with the Word.*"

> *One of the first persons who offered himself to my worship was a descendant of Thales, called Telliamed, who told me that mountains and men were produced by the waters of the sea. At first, there were handsome seamen who later became amphibious. Their beautiful forked tails changed into thighs and legs. I was still permeated by Ovid's Metamorphoses and by a book demonstrating that the human race was the bastard of a race of baboons. I much preferred being a descendant of a fish than a monkey.*[23]

But the character in Voltaire's book ends up by having doubts about the importance of the actions of the sea:

> *But, incredulous Sir, what comment have you to make about the petrified oysters found on the summit of the Alps?*
>
> *- I shall answer, Sir creator, that I have seen neither petrified oysters nor vessel anchors at the top of mount Cenis. I shall answer what has already been said, that oyster shells (which petrify easily) have been found at very large distances from the sea, just as Roman medals have been unearthed at a hundred leagues from Rome. And I prefer to believe that pilgrims of Saint-James have left a few shells near Saint-Maurice than to imagine that the sea made Mount Saint-Bernard.*

There follows a long diatribe about experiments on spontaneous generation.

*The History of the Earth* by Georges Louis Leclerc, count of Buffon (1707-1788), is a much more important landmark in the history of biology than is the *Telliamed.* (The first part was published in 1744.) It is part of Buffon's gigantic *Natural History*, an opus of 44 volumes with hundreds of hand-coloured plates which took 55 years to complete. The work reflects the 18th century scientific and encyclopaedic ambition, partly inherited from Linnaeus, to put all knowledge on nature into some kind of order. The writings of Buffon, who at the time was Keeper of the King's Garden (which was later to become the Natural History Museum), were quite daring. Buffon thought that all planets were initially part of the same celestial body of which the sun was the central core. Earth's matter had been detached from the sun by the collision of a passing comet. On cooling, the earth blistered and gave rise to mountain chains, and the atmosphere condensed into a single ocean. Fossils and sedimentary rocks date from this time. As the waters receded, continents emerged and the reign of plants and of the great terrestrial animals began. The earth was still tepid; tropical animals probably still inhabited areas which are now temperate or cold. Finally, today's species emerged and, among them, the human species. Buffon even suggested that man and monkey might belong to the same family and have a common ancestry. He went as far as to say that the monkey was a degenerate human.

Buffon's book lays the foundations of the theory of actual causes. Nature is seen as a set of causal relationships. The changes undergone by the earth's surface result from constant and gradual physical forces that are still operating today and that are therefore in no way exceptional. This theory will be developed later by Charles Lyell in *Principles of Geology* (1830) but, at the time, it provoked strong reactions as reported by the Marquess of Argenson:

> *Sire Buffon's head is in a spin from the worry caused by his book's success. The devout are furious and want it to be burnt by the executioner. Verily, he contradicts the Bible in all points.*[24]

On the 15th January 1751, the Theology Faculty of the Sorbonne (whose judgement was backed by Voltaire, albeit an atheist) condemned 16 statements by Buffon and demanded that he withdraw them. Buffon obliged. In the next volume of *Natural History*, he wrote:

> *I declare that I had no intention of contradicting the Scriptures, that I believe most firmly all that is stated therein about the Creation [...] I retract everything in my book about the Earth's formation and, in general, all that may be contrary to the narrative of Moses.*[25]

A highly controversial issue was the estimated age of the Earth. In 1650, archbishop Ussher had put the Earth's origin at 4004 years B.C. In 1744, in *Theory of the Earth*, Buffon proposed that this date be pushed back by about 74,000 years. Was his retraction genuine, at least on this point? Hardly so, because his unpublished notes give ages greater than 500,000 years. He confided to Hérault de Séchelles:

> *The people need a religion [....]. When the Sorbonne took quarrel with me, I gave it all the satisfaction it asked. This is mere tomfoolery; men are so stupid that this satisfies them.* [26]

Despite Buffon's quarrels with the Sorbonne, his reputation as a dilettante, and his grand airs (he liked to say that there were only five truly great men: Newton, Bacon, Leibniz, Montesquieu ... and himself), his view of the time-scale of the Earth's history was a real eye-opener to contemporary intellectual and cultural society.

**How Buffon calculated the age of the Earth**

According to Buffon's hypothesis, the Earth was born because a comet hit the Sun at an oblique angle and smashed away about one part in 650. The liquid mass then divided into several rapidly rotating spheres. The resultant centrifugal force ejected part of the matter of the protoplanets and is at the origin of the satellites of our Solar System's planets.

Buffon performed a series of experiments on the cooling times of iron balls of different size. He extrapolated his results to calculate *"with Newton, how long it would take for a sphere of the size of the Earth to cool"*.[27] He found that it would take 42,964 years and 211 days before such a sphere could be touched without scorching, and 96,670 years and 132 days to reach the present temperature. But the Earth is not only made of iron, and Buffon repeated his cooling experiments with other constituents. Finally, discarding all but *"glass, clay, hard sandstone, marble and ferrous materials"*, he obtained the following results: 2,095 years to consolidate, 33,911 years before touching, and 74,047 years to reach the present temperature or rather 74,832 years if due account is taken of the Sun's heat whilst the Earth cools.

However, after examining sedimentary deposits on the Normandy coast, Buffon decided that his cooling period was too short and looked for hidden causes that might delay cooling. *"He tested various hypotheses which yielded increasing ages, from 700,000 to 800,000 years for the primitive ocean to stabilise and one million years for the present age of the Earth. Then, respectively, two and three million years, which multiplies by 40 the action of the hidden causes. This coefficient was no doubt too small, and the Earth should be assigned an age of at least ten million years"*.[28]

However, in *Epochs of Nature* (1779), he will publish the estimate of 75,000 years in order, as he said, not to plunge his readers into *"the dark depths of time"*. *"It is true that the further we extend time, the nearer we shall be to the truth and to Nature's use of time. Nevertheless, we must shorten it as much as possible so that it is in keeping with the limited power of our intellect"*.[28]

## The precursors of transformism

With hindsight, it seems that several conditions had to be met for the breach made by Buffon to open up onto genuine evolutionism:

- People's minds had to be emancipated from the traditions of the bible and the hold of theologians. Buffon was a good example indeed! In particular, the use of words such as Creator, Supreme Being had to be abandoned for the simpler use of the word Nature.

- One had to recognise how old is the World in order to explain the discovery of the remains of extinct species and the transformation indices of the Earth.
- One had also to accept that present species could change beyond the limits set by fixism, even if fixism was a most valuable framework for the naturalist. All life, like mankind, is part of history, a history where forms that are separate today may have been linked in times past.

Strictly speaking, Charles Bonnet (1720-1793), a naturalist from Geneva, was not a precursor of transformism. We mention him because he introduced the word evolution to denote the process of the growth of an individual from seed. Bonnet, in fact, belonged to the preformationist school.[29] At the time, the notion of a novel structure emerging from a primitive one was as difficult to accept as was the idea of transformism or evolution.

**The doctrine of preformation**

Jan Swammerdam (1637-1680) was the originator of the doctrine of preformation. This doctrine rejects procreation since parents cannot create a being that is already perfectly formed. God initially created the seeds of all subsequent offspring; they are nested inside each other. The beginning of genital life is just the "awakening" of a seed (whether from the father or mother) which sets out on the road of "evolution", growth and development from its initial numb state. All those who believed that living beings come from an egg, or that the embryo is either partly or totally preformed in the little animal that is the sperm, espoused this doctrine. Marcello Malpighi (1628-1694) thought that there was a pre-existent seed (which in his study was the unfertilised egg of the chicken) but which, however, harboured only part of the individual. As pointed out by Bernardin de Saint Pierre in 1797, this doctrine was more of a political and social ideology rather than a genuine scientific conviction: *"It is upon this opinion, so often refuted by experiment, that the aristocracy founds its prerogatives. It is supported by the subtle reasoning of our schools, which flatter all tyrannies. They say that all men, from fathers to sons, were present in the first man, like a pile of goblets."* [30]

Although Bonnet was not an ardent defender of species variability, he apparently did not reject the idea. He compared the growth of the individual with the history of the species. He wrote of the chicken embryo:

*The phases through which we see it pass enable us to assess the upheavals that organised bodies have had to undergo to reach their latest form [and help us] imagine the new forms that animals will adopt in their future state.*[31]

Bonnet even predicted that man will improve so much that he will become an angel and depart from an Earth no longer suited to this new status!

The work of Pierre Moreau de Maupertuis (1698-1759), in particular *Physical Venus* (1754), is a cornerstone in the history of transformism and evolutionism. Maupertuis not only thought transmutation and diversification of species possible but tried to explain them by processes from embryology. He was especially interested in heredity. Having seen a "white negro" (albino), he concluded that being white was merely a hereditary trait. It was just like polydactily or like other features he had encountered in the breeding of poultry and dogs. His intuition was astonishing. For example, he guessed that heredity results from pairs of what he called "molecules" from the father and mother. These have affinity for each other and combine in the foetus. Sometimes, this process goes awry and produces an unexpected being which may be at the origin of a new species. He wrote:

*Could we not thus explain how just two individuals have given rise to a multiplicity of highly dissimilar species? They would initially have arisen from fortuitous productions in which the elementary parts were less orderly than in the father and mother animals. Each erroneous step would have led to a new species; the infinitely diverse animals we see today would have resulted from such repeated errors.*[32]

The philosopher Denis Diderot (1713-1784) was not quite of the same opinion as either Bonnet or Maupertuis but he also subscribed to evolutionist metaphysics, whilst adding a kind of dynamic materialism. He vindicated the unity of nature in the organic world; he adopted Buffon's idea of a chain of beings and thus implicitly criticised Linnean nomenclature. He introduced, however, a dynamic dimension allowing for the existence of a unique act in nature or of a prototype of all beings. Diderot unfolded these ideas in *Interpretation of Nature* and in *D'Alembert's Dream.* In these works, he focused on the consequences of use and non-use.

**Diderot's interpretation of nature**

*Surprise often arises because we assume there are several prodigies when there is only one; because we imagine that in nature there are as many specific actions as there are phenomena, whereas nature may have produced but a single action. It would even seem that if nature had had to produce several actions, their results would have been kept isolated; that there would have been collections of phenomena that are independent of each other; and that the general chain resting on a philosophy of continuity would be broken in several places. The absolute independence of a single event is incompatible with the idea of a whole; and without the notion of a whole, there can be no philosophy.*

*It seems as if nature has taken pleasure in making variations on a single mechanism in an infinite variety of ways. It abandons a given mode of production only once it has multiplied individuals in all manner of ways. When we observe the animal kingdom and note that, amongst four-legged animals, there is not one that does not have functions and parts - especially inner parts - that are totally similar to those of another four-legged animal, then we have to believe that there was a first animal, a prototype for all animals, in which nature has but lengthened, shortened, transformed, multiplied, and abolished certain organs.*[33]

*Doctor Bordeu. - Organs produce needs and, conversely, needs produce organs.*

*Mademoiselle de Lespinasse. - Doctor, are you delirious?*

*Bordeu. - Why not? I have seen two stumps become two arms.*

*Mademoiselle de Lespinasse. - You're lying.*

*Bordeu. - It's true; but instead of two missing arms, I have seen two shoulder blades lengthen, move like claws, and become two stumps.*

*Mademoiselle de Lespinasse. - What folly!*

*Bordeu. - It's fact. Imagine a series of armless generations - imagine continuous efforts - and you will see the two sides of a small claw extend themselves, further and further, cross over in the back, come round to the front - maybe become digitated at their extremity, and recreate arms and hands. The original conformation worsens or improves through need and habitual use. We walk so little, work so little, and we think so much that I have not lost hope that man will end up by being just a head.*

*Mademoiselle de Lespinasse. - A head! A mere head; I trust that unrestrained politeness... You are giving me quite ridiculous ideas.*[34]

Yet again, scientific and philosophical ideas evolved in line. "Time-scaling" of the chain of life, a feature of 18th century thinking, led to the notion of gradual progress. Henceforth, it was accepted that the rungs of life's ladder had appeared one after another, the lower forms first; there was no absolute limit to improvement. This train of thought was also applied to human societies, to their cultural, political, legal, and religious foundations. But might the idea that the world evolves not shake these foundations? What status has a mankind known to be related to apes? Of course, the intellectual and moral authorities of the time refused to ask such radical questions but these issues nevertheless infiltrated people's minds.

The similarities among organs studied by Buffon and Daubenton strengthened the idea of a teleological scheme of life. The emergence and arrangement of shapes and processes were determined by a plan that was to be fulfilled, or a goal that had to be met. Jean-Baptiste Robinet (1735-1820) and Johann Gottfried Herder (1744-1803) thought that man might be the end product of a prototype that was present in all living beings and that had undergone gradual improvement.

**J.G. Herder's view of man**

*If we may hazard a guess about the deep mystery as to where creation took place, we could consider the human genus as the great confluent of inferior organic forces that merged to form mankind. But what happened next? Man has brought God's image with him and benefited from the most delicate organisation that the earth could offer; must he turn round and revert to a trunk, plant ,or elephant? Or has the wheel of creation stopped with him?... Let us look back and examine how, behind us, everything seems to have matured in order to make man, and how all that is new in man is the first bud and seed of what he must be and for what he is destined.*[35]

According to Herder, an "active force" or a divine power had produced a dynamic hierarchy of forms in which mankind had its place but this view did not exclude the spontaneous generation of species nor a series of creations. Its rationale was not so much transformism as a kind of naturalist romanticism which, under the influence of the Stoic tradition, took account of the passing of time.

Admittedly, these theories, whether they were scientific or philosophical, were still imprecise and lacked convincing arguments. Voltaire, as we have seen, had no trouble in ridiculing them. He lampooned Maupertuis in *Doctor Akakia's Life Story* (1752-1753) by mingling medicine, embryogenesis, and the origin of the soul... with an excavation project which reaches the centre of the earth!

*We suggest that the next time the young author [of the Vénus Physique] is in the throes of the act of procreation with his wife (if he has one), he should no longer think that the child is formed in the uterus by attraction; and we urge him, if he commits the sin of the flesh, not to envy the fate of the love-making snail or that of the toad, and to imitate less the style of Fontenelle once maturity will have shaped his own.*[36]

In the natural sciences, the chief adversary of transformists was the eminent paleontologist Georges Cuvier (1769-1832) who doggedly defended fixism in biology and promoted catastrophism in geology. Nicknamed the "dictator of biology", Cuvier was above all the founder of modern comparative zoology. He earned this title not because he adduced new facts but because he introduced a new way of observing living organisms. According to Cuvier, the organs of an animal are functionally interrelated. Thus, a carnivore must have claws and teeth but also good eyes, a system able to digest meat, and quick reflexes. Vicq d'Azyr had already postulated a theory of correlations but Cuvier turned it into a foundation for a classification system. He applied it to the study of fossils, especially those found around Paris, in order to reconstruct the forms of past creatures.

By throwing a bridge between his two fields of work (classifying living creatures and reconstructing animals from fossils), Cuvier was able to revise the prevailing linear and hierarchical view of life's ladder which compared creatures of different degrees of perfection. He proposed instead that the animal kingdom be divided into four groups ("*embranchements*") for valid comparisons within each group: Vertebrata, Mollusca, Articulata, and Radiata.

Cuvier's explanation for the existence of fossils of extinct species was catastrophism. Sudden, violent catastrophes had destroyed the species whose fossilised remains we now find, leaving room for new fauna; a series of "creations" and "catastrophes" explained the appearance and disappearance of highly diverse forms of life. This thesis neither brought into question the dogma of the fixity of species, nor called for the existence of intermediate forms. Alcide D'Orbigny (1802-1857), a disciple of Cuvier, imagined that as many as 27 catastrophes had taken place! Such lack of moderation in combating the idea of evolution prepared the ground for a genuine evolutionary theory. The theory arose despite the intellectual dictatorship and renown of Cuvier and thanks to the fight put up by his colleague Lamarck at the Natural History Museum. Cuvier, however, cannot be dismissed without mentioning the famous controversy that opposed him in the early 1830s to Etienne Geoffroy Saint-Hilaire (1792-1844).

When, in 1795, Geoffroy Saint-Hilaire invited Cuvier to Paris and obtained him a position at the Natural History Museum, he little suspected that this man would become a fearsome rival. Although Geoffroy did do some observational work, chiefly during the scientific expedition organised by Bonaparte in Egypt, he was above all attracted by the world of ideas. He proposed three principles for deciphering the living world: unity of organic composition ("unity of plan") (nature has created all living things according to a unique plan with many variations), connections (organs with a common embryologic origin may display different forms and functions depending upon the organism), organ balance ("*No normal or pathological organ ever thrives exceptionally without another organ of the same or related system suffering accordingly*", 1822).

One reason for the controversy that opposed the two men was the "crocodile affair". Cuvier had begun studying the fossils of crocodiles found near Caen and Honfleur in France. (They were, in fact, gavials, fine-jawed crocodiles that are nowadays found in India). Geoffroy, in a report criticising Cuvier's analysis of the fossils, upheld the thesis of the evolution of species, claiming that there had not been a series of creations in life but a continuity. The full title of his report is noteworthy: *Research into the organisation of Gavials. On the subject of their natural affinities which result in the need for another generic distribution, Gavialis, Teleosaurus and Steneosaurus; and on the question whether gavials (Gavialis), now found in the eastern parts of Asia, are descended, by an uninterrupted lineage, from antidiluvian Gavials, either the fossil Gavials called Caen crocodiles (Teleosaurus) or the fossil Gavials of Le Havre and Honfleur (Steneosaurus)* (1825).

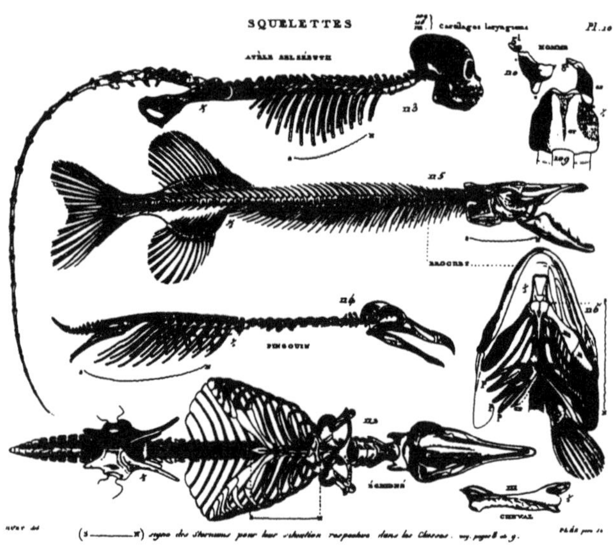


*Figure 1.2.* The concept of an organisation plan. Plate from Anatomical Philosophy (1818) by Etienne Geoffroy Saint-Hilaire.

Up to this time, Geoffroy had not believed in Lamarck's statements on transformism. To him, the controversy surrounding the crocodile business was a eye-opener even though he subscribed, for the most part, to a view of evolution that was largely inherited from embryology, one that was limited to the unfolding of a pre-established plan.

The debate focused on whether there was a single plan for the entire animal kingdom. Could the plan, which seemed acceptable for Mammals and even Vertebrates, be extended to Insects and Molluscs? In 1830, during a sitting of the Academy of Sciences, Cuvier challenged Geoffroy's hypothesis and tried to make out that his examples were mere analogies, like those already observed by Aristotle. But Geoffroy Saint-Hilaire put up a fight and presented a paper entitled: *On the theory of analogies, to establish its novelty as a doctrine and its practical use as a tool.* The "quarrel of analogies" had begun.

**What does the word analogy mean in biology?**

Analogy denotes a functional similarity in structures of different origin. We thus speak of the analogy between the wing of an insect and of a bird. Analogy is opposed to homology. Structures are homologous if their build is based on the same morphogenetic model, whether they fulfil the same function or not. Thus the front limbs of the shrew-mouse, mole, horse, bat and man are identical in structure, bone for bone, but their function varies from searching to running, from flying to typing on a computer keyboard!

Despite the clear-cut difference in the definitions of the two terms, the distinction has never really been upheld because species changes over time concern both form and function. Richard Owen, the English paleontologist who coined the term homology in 1843 used it to denote the same organ in different animals with all possible variations in *form* and *function*. "*The existence of homologies,* concludes Charles Devillers, *is an unavoidable consequence of evolutionary transformation.*" [37] The use of the terms is nevertheless ambiguous as it is never easy to account for the structure and function of living things. The controversy between Geoffroy Saint-Hilaire and Cuvier is an illustration. If Geoffroy had analysed the notions of analogy and homology in greater depth (he used the latter notion in 1825 but Owen consecrated the use of the term), he may not have made the mistake of relating the cuticle of insects to the vertebra of vertebrates.

*Figure 1.3.* The wings of a bird (top) and a bat (middle) are homologous. They are derived from the same embryonic tissues and have the same bones and same connections. On the other hand, the wing of an insect (bottom) is not derived from the same structures during development. It is analogous, and not homologous, to the other wings.

Much was written in the wake of the quarrel because, as pointed out by Hervé Le Guyader, it was no easy task to apportion right and wrong:

> *Was Cuvier right in challenging Geoffroy about the existence of a unique organisation plan? Yes, without any doubt. Insects such as "dermo-vertebrates" and the diaphragm of the squid are frightening suggestions. Was Geoffroy right in postulating a unique plan? Yes, without any doubt but not in the way he did! Thus Geoffroy was right in being wrong and Cuvier was wrong in being right!* [38]

When Cuvier claimed that there were four basic organisation plans, he was not defending fixism but simply rejecting a universal analogy that he did not think was founded on fact. And when Geoffroy thought that all beings were part of a unique plan, he was not embracing transformism but simply using the old concept of life's ladder, versions of which had been used, even abused, by 18th century naturalists.

There is much to learn from such a controversy as Goethe was among the first to point out. His analysis of the arguments of both protagonists showed that most of the terms they used had double meanings. However, as H. Le Guyader rightly concludes, "*was it not this ambiguity that permitted the development of Cuvier's and Geoffroy's ideas?*"

# 3. LAMARCK, FOUNDING FATHER OF THE THEORY OF TRANSFORMISM

*Of the three authors whom we like to think are at the origin of modern biology, Lamarck is, without doubt, the one whom history has treated least well. The works of Claude Bernard and Charles Darwin may have been somewhat debased, often in the name of scientism, but Lamarck has been almost slandered and used as a foil by scientists and philosophers holding all kinds of opinions.*[39]

There are many reasons why such disrepute should have fallen upon the shoulders of Jean-Baptiste de Monet, Chevalier de Lamarck (1744-1829), adds the historian of science André Pichot. An obvious one is the timing of his major work, *Zoological Philosophy,* which was published in 1809 when he was already 65 years old. Lamarck had been tutor to Buffon's son and was, first and foremost, a man of the 18th century, of the Enlightenment, who professed a materialistic philosophy and revolutionary political ideas. However, this line of thinking had been disavowed by the Empire and the Restoration. The beginning of the 19th century set a kind of spiritualism against 18th century materialism, so much so that Lamarck was subsequently taxed with transcendentalism, an opinion which positivist scientific thinking, however, rejects later on. One must also mention that Cuvier dominated the biology of the time, holding the chair of Vertebrates at the Museum, whereas Lamarck obtained that of Insects and Worms in 1793. Lastly, Lamarck's papers in physics, chemistry, pneumatic theory, the matter of fire and sound, or hydrogeology were worthless and behind the times. He was against modern chemistry, against Lavoisier, defended phlogiston theory and published *Meteorological Directories* that made predictions which did little to enhance his reputation.

**Phlogiston theory**

The term "*phlogistique*" appeared in the French language in 1747 and was used in chemistry in the second half of the 18th century. It denotes the fluid presumed to be present in all bodies and which, when leaving a body, results in its combustion. Fire was in fact considered to be one of the materials, or principles, that compose bodies. Phlogiston theory dominated for 50 years. It was finally refuted by Lavoisier.

The transformist ideas that Lamarck developed in his fifties and expressed in his *Opening Addresses* of Year VIII (1800), Year X (1802), Year XI (1803) and in 1806 - and chiefly in *Zoological Philosophy* - were not really novel. We have already mentioned several of his precursors (Buffon, Bonnet, Robinet). However, Lamarck's work was not a mere compendium of ideas but a coherent, systematic entity that makes him the true founding father of transformism.

## The order of nature

The origin of Lamarck's transformist ideas probably lies in his classification work, mainly on Invertebrates. Classification was not an easy task because it depended on defining the notion of species and establishing which forms in systematics belonged to the same or different groups. Lamarck began his career as a botanist. He knew that the artificial boundaries of plant systematics were often violated, plant forms being highly variable and sensitive to culture conditions. This led him to defend a nominalist view of systematics units:

> *We can rest assured that, among its productions, nature has not created classes, orders, families, genera, or constant species, but only individuals that succeed each other and that look like those that produced them.*[40]

Lamarck revived the idea of continuity among living beings that had been contested because of the gaps. These gaps were now being filled by the discoveries of naturalists. The picture was getting clearer and clearer even though taxonomists were not yet out of a job:

> *Fortunately for the practice of the art we need to introduce into our classifications, there are so many races of animals and plants that are still unknown to us - and so many that will probably remain unknown to us because of the obstacles that are the places they inhabit and other circumstances - that the gaps which arise, whether in a series of animals or plants, will for a long time still and perhaps for ever provide us with the means of setting up most of the lines of division.*[41]

Lamarck's nominalistic approach thus does not lead to rejection of all natural order but replaces the old order founded on resemblance by a new order founded on the genealogy of species. We could summarise the history of systematics by this transition from a criterion of resemblance to one of affiliation.[42]

**Genealogy, mutability and increasing complexity of species according to Lamarck**

> *The belief is now fully justified that nature has established an order among her productions within each kingdom of living bodies. It is the order in which each of these bodies was originally formed [...].*

*This order, as we shall get to know it, will replace the systematic or artificial classifications that we have been forced to create in order to arrange conveniently the different natural bodies that we have observed.*[43]

*Of Species and of the meaning we should attach to this word. It is untrue that species are as old as nature, and that they have all existed as long as each other. But it is true that species were formed one after another, that they exhibit only relative constancy, and that they are only temporarily unchanging.*[44]

*In her workings, nature began, and still begins every day, by fashioning the simplest of organised bodies and [...] it is only these bodies that she fashions directly, that is to say the rudiments of organisation covered by the term spontaneous generation [...]. Because each body part's ability to grow is an early manifestation of life, it gave rise to different modes of multiplication and regeneration of individuals; improvements in the way the parts are organised, and in the shape and variety of the parts, have thus been preserved.*[45]

As species succeed each other with time, Lamarck asked whether it was right to classify forms from the most to the least accomplished, that is from mammals to infusoria. It was the practice to give preference to that which appealed, pleased or interested, to go from the best to the least known. But Lamarck thought this direction was contrary "*to the very order of nature*". Nature starts off with the most imperfect animals and those with the simplest organisation and ends up with the most accomplished because, having made them one after another, "*she must have started with the simplest, and only in the end produced those with the most composite organisation*". Lamarck thus acknowledges complexity as a criterion but applies it within a historical framework. The degree of complexity of an organisation is an index of time elapsed. The title and summary of the sixth chapter of the first part of *Zoological Philosophy* summarise this idea of a dual - synchronic and diachronic - sequence:

*Degradation and simplification of the organisation from one end of the animal chain to the other, starting from the most composite and moving toward the simplest. It is a positive fact that when following, as is customary, the animal chain from the most perfect to the most imperfect, one observes increased degradation and simplification in organisation. Consequently, when moving along the animal ladder in the opposite direction, in nature's order, one will find increasing composition in the organisation of animals, a composition that will have a delicate and regular progression as long as the circumstances of habitat, lifestyle, etc... have not induced various kinds of abnormalities.*[46]

According to Lamarck's conception of natural order, primitive and evolved species must co-exist. All species undergo a gradual transformation process, but they have reached different stages of this process. The simplest beings are being renewed continuously. We should therefore find in nature today all organisational stages with the corresponding species. How then does one explain the existence of fossils of lost species? Shouldn't we be surprised "*if amongst the many fossils we find in all the dry areas of the Earth, and that offer us the debris of so many animals that once existed, there are so few for which we recognise living analogues?*"[47] "*Paradoxically*, notes Pichot, *fossils, which we usually consider to be paleontological evidence for evolution, are impediments in Lamarck's transformist thesis*".[48] If these species have disappeared from nature, Lamarck finally concludes, it is not because they were eradicated by the first human beings but because they have been modified into other close forms under the influence of environmental changes which induce "*changes in needs, habits and lifestyle*". These changes, which supplement the process of increasing complexity that always leads to the appearance of the same types, yield a wide variety of forms:

> *These irregular variations in the improvement and degradation of non-essential organs occur because these organs, more than any others, are subject to the influence of external circumstances; these circumstances also induce similar variations in the form and state of the most external parts, and give rise to a diversity of species that is so great and so singularly ordered that instead of being able to arrange them in a unique, simple and linear series, according to a regular, graduated scale, these same species often form, around the masses to which they belong, lateral branches whose extremities are truly isolated points.*

Lamarck's transformist explanation for "lost species" clearly contradicts Cuvier's catastrophe theory which he saw as an *ad hoc* argument to fit the situation. His hypothesis is based, on the contrary, on a slow and gradual action of nature which he considers to be more probable.

> *But why are we to assume a universal catastrophe without any proof when just a better understanding of the way nature proceeds can account for the observations which we make on all of its aspects?*
>
> *If we consider, on the one hand, that whatever nature does, it does not do abruptly, and that it acts everywhere slowly and in stages, and, on the other hand, that the specific or local causes of disorders, upheavals, shifts, etc - though subject to the laws and progress of nature - can account for everything that we observe on the surface of the earth, then*

*we shall realise that it is quite unnecessary to assume that a universal catastrophe came to upset and destroy a large part of nature's own works.*[49]

Lamarck's linear view of life was not accepted by several biologists of the beginning of the 19th century. We have already mentioned Georges Cuvier, but we should also mention Karl Ernst von Baer (1792-1876) and Louis Agassiz (1807-1873). Like most naturalists of the time, they classified animals into distinct basic types[50] which could in no way have arisen by gradual evolution of a unique prototype. Exploiting his observations in embryology (those on the chicken are known best), von Baer rejected the Meckel-Serres law, or law of embryologic parallelism, which stated that, during their development, higher-order animals reproduce the structures of lower-order animals. According to von Baer, the individual develops within limits that are set for each type; the type of organisation determines the mode of development. This meant that embryology could be of decisive help to comparative anatomy, and that both evolutionism and progressism were rejected. (As a result, these two notions were all the more readily confused later on).

At this time, the tree of life, a sequel to the chain of being, offered a new paradigm for biology. It took Cuvier and von Baer's criticisms into account without however rejecting evolutionism. If all animals are not derived from a single chain of life, there was no need to imagine a unity of plan.

## The principles of transformism

An extract from the *Natural History of Animals without Vertebrae* will illustrate Lamarck's understanding of nature's slow and gradual workings over time:

*Nature's plan of action with respect to the production of animals is clear from the prime and predominant cause that confers animal life the ability to gradually compose and improve its organisation, not only its overall organisation, but each specific organ system, as and when they are established. This plan - this gradually composed organisation - has been carried out, by this prime cause, in existing animals.*

*But an external, accidental, and therefore variable cause, has here and there cut across the execution of this plan without destroying it, as I shall prove. This cause has, in effect, given rise to genuine gaps in the series, to finite branches that shoot off from various points and alter its simplicity, or even to abnormalities that we observe in the specific organ systems of these organisations.*[51]

Lamarck did not provide a clear explanation for the hereditary process underlying the trend toward increasing complexity.[52] He only mentioned the trend to offset it against the circumstances that will upset its linearity, induce species to diversify, and disturb natural order. Natural order is only established incompletely, and only by means of the opposition it encounters.

Lamarck based his model on two laws, the modification of the organ by modification of needs and habits and the inheritance of acquired characters:

> *First law. In any animal that has not reached the end of its development, a more frequent and sustained use of any organ gradually strengthens, develops and enlarges that organ, and confers it a power proportional to the period of use, whereas the constant disuse of any organ imperceptibly weakens it, leads to its deterioration, gradually decreases its functional capacity, and ends up in its disappearance.*
>
> *Second law. Everything that nature has helped individuals acquire or lose through the influence of the conditions to which their race has long been exposed and, hence, through the influence of the predominant use or permanent disuse of an organ, is conserved by the new individuals that are engendered provided that the acquired changes are common to both sexes, or at least to the genitors of these new individuals.*[53]

The examples Lamarck gave to illustrate his first law are famous. The giraffe has an extremely long neck because it munches the leaves of trees, ducks' palms come from the increased space between digits, snakes' feet have regressed because snakes have the habit of crawling.[54] Why did Lamarck insist so much on needs rather than on organs? Probably because he wished to dissociate himself from any statement on the divine creation of beings (with or without the intervention of catastrophes) even if, in both his mind and that of Erasmus Darwin, grandfather to Charles Darwin, transformism and deism were not really mutually exclusive. When God created the Universe, He incorporated evolution as a specific means of accomplishing His project for the living world and did not resort to miracles. Even if Lamarck used arguments that were quasi-scientific whereas Erasmus Darwin held a more schematic, even poetic, reasoning, both scientists really wanted to propose a genuine "philosophy of nature" rather than just biological hypotheses.[55]

**Erasmus Darwin**

Erasmus Darwin (1731-1802) was an eminent supporter of the industrial revolution in 18th century England. He founded the *Lunar Society* which fostered the exchange of scientific, technical and cultural information. He was a fierce opponent of slavery, defended the French revolution and democracy, and kept company with Benjamin Franklin and Jean-Jacques Rousseau. His most important work is the two-volume *Zoonomia: Or The Laws of Organic Life* (1794-1796), a treatise in medicine that addresses innate and acquired behaviour as well as evolutionism. The great variety of forms in nature (his book is as rich in observations as are the works of his famous grandson) would be due to a triad of needs (sex, food, safety) that transform living beings over generations. Like Buffon and Needham, Erasmus Darwin believed in the spontaneous generation of lower organisms which subsequently undergo improvements. He was, nevertheless, struck by the great difference between the state of inorganic and organic matter. Reproved by the Anglican canon William Paley, author of *Natural Theology or Evidences of the Existence and Attributes of the Deity Collected from the Appearances of Nature* (1802), *Zoonomia* was condemned by the Sacred Congregation of the Index of the Catholic church.

The hypothesis that acquired characters are inherited is customarily linked with Lamarckism but this calls for several comments. First, Lamarck himself did not use the expression simply because, as Pichot notes:

> *there was no clear opposition between what was innate and what was acquired. It was known that an individual received certain characters at birth and acquired others but the distinction went no further. Once these inherited and acquired characters were "mixed" in an individual, they were no longer separate, and the transmission to the offspring concerned as much the one as the other.*[56]

More broadly, are not all characters more or less acquired? The phrase - transmitting the results of training to future generations - is more in keeping with Lamarck's thinking. But the idea was not exclusively his. At the time, it was so widely accepted that he saw no point in discussing it in *Zoological Philosophy*. Aristotle had already mentioned it,[57] Darwin will propose an explanatory model, and it will not be abandoned (for ever?) until the end of the 19th century and the work of August Weismann. Thus we do not owe the law of acquired characters to Lamarck even if he did give it more emphasis than his contemporaries did and, in so doing, made it a major limitation of his transformist theory.

## From Quadrumana to Bimana?

Linnaeus had introduced man into his classification, thus linking him to animals and in particular to apes. The church as well as scientists such as

Buffon had opposed this move, calling chiefly upon the fundamental separation between body and soul which confers human beings with a special status. Lamarck went further than Linnaeus because, according to his transformist hypothesis, man descends from the ape. He chose his statements with care, used the conditional ("if", "let us now suppose"), but whom was he fooling?

**From ape to man (Lamarck)**

*If man were distinguished from animals by his organisation only, it would be easy to show that the organisational characters we use to form a separate family with its varieties are all the products of past changes in his actions and of the habits he has acquired and that have become specific to the individuals of his species.*

*Indeed, if a race of quadrumana, especially the most perfect one, were to lose, by force of circumstance or for some other reason, the habit of climbing trees and grasping branches with its hands in order to cling to them; and if the individuals of this race were forced over a series of generations to use their feet just for walking and stopped using their hands as feet; there is no doubt, according to the observations described in the previous chapter, that these quadrumana would at length have become bimana and that their big toes would have ceased parting from their other toes as their feet would be used for walking only.*

*Moreover, if the individuals of which I speak, driven by a need to dominate and to see both far and wide, were to endeavour to stand upright and adopted this stance for generations, there is no doubt either that their feet would imperceptibly acquire a shape suitable for standing erect, that their legs would acquire calves, and that these animals would only be able to walk with difficulty on both hands and feet.*

*Finally, if these individuals were to stop using their jaws as weapons with which to bite, tear or grasp, or as nippers to cut and feed on grass, and if they were to use them just to chew, there is no doubt either that their facial angle would open, that their snout would become shorter and shorter and finally disappear, leaving their incisors vertical.*

*Let us now suppose that a race of quadrumana, say the most accomplished, acquired through unchanging habits among all its individuals the conformation that I have just described, and the faculty of standing and walking upright, and that it ended up gaining supremacy over other animal races, we can then easily conceive:*

> *1. That this race, having subdued other races thanks to its highly advanced faculties, would take possession of all the sites on the earth's surface that were suited to it;*
>
> *2. That it would drive out any other eminent races able to dispute the goods of the earth, and force them to take refuge in sites it did not occupy;*
>
> *3. That it would hinder the multiplication of neighbouring races and relegate them to woods and other deserted sites, thereby arresting the development of their faculties toward perfection; whereas being able itself to spread everywhere, to multiply without hindrance, and to live in numerous hordes, it would create successively new needs which stimulated its own activity and gradually improved its skills and faculties;*
>
> *4. Finally, that this predominant race, having acquired absolute supremacy over all others, would manage to establish a difference, and set a considerable distance, between itself and the most accomplished animals.*
>
> *The most consummate race of quadramana could thus have become dominant; have changed its habits as a result of the absolute ascendancy it exercised over others, and of its new wants; have undergone gradual changes in its organisation and acquired many new faculties; have kept the most accomplished of the other races in check in the state they had reached; and have wrought very striking differences between these races and itself.* [58]

In *The Origin of Species* Charles Darwin did not dare go as far as Lamarck but he will make up for it in *The Descent of Man*. Lamarck deserves this long citation as a homage because, once we turn to the Darwinian paradigm, he will have to fall into oblivion. French culture, however, has perpetuated his memory. To illustrate how Lamarck has left his mark, we only need to refer to Emile Zola's work. Zola's naturalism and his curiosity about heredity is plain to see as soon as we read the preface to the first volume of Les Rougon-Macquart (1871):

> *I want to explain how a family, a small group of people, behave in society, how they prosper and give rise to ten, twenty individuals who, at first sight, seem so different but, on analysis, are found to be closely linked to each other. Like gravity, heredity has its laws.*
>
> *I shall try to find the thread that leads mathematically from one man to the next, and then follow it. When I will have gathered all the threads together and will hold a social group in my hands, I shall show how the group goes about its business and the role it plays during an important*

*period in history. I shall make it struggle in all its complexity, I shall analyse the will of each member and the overall thrust of the whole [...]*

*In my mind, this work, which will be made up of several episodes, is the natural and social history of a family during the Second Empire. The scientific title of the first episode (The Fortune of the Rougons) is The Origins.*[59]

In *Le Doctor Pascal* (1892), which summarises and concludes his work, Zola published the family tree of the Rougon-Macquart. The tree is a field of observation for the doctor, a student of heredity

*who had but one belief; he believed in life. Life was the only divine manifestation. Life was God, the great driving force, the soul of the universe. And life had no other instrument but heredity; heredity made the world, so that if we could understand and capture heredity in order to do with it according to our will, then we would have made the world as we wished it to be.*[60]

The hereditarism, that leads Pascal-Zola toadmit that there are people in the community who are a danger to their descendants, is counterbalanced by his conviction that life itself holds a regenerative seed and that the environment can modify the nature of individuals. Clotilde Rougon, Pascal's niece, his assistant and lover, expresses this conviction when, in the last pages of *Les Rougon-Macquart*, she contemplates the family tree:

*Clotilde had taken an involuntary glance at the family tree which lay spread out before her. The threat was there, so many crimes, so much mud, amidst so many tears and so much good and suffering! Such an extraordinary mixture of the best and the worst, a short-version of mankind with all its failures and struggles! Would it not have been better if a thunderbolt had wiped out this desecrated and miserable ants-nest? After so many terrible Rougons and so many abominable Macquarts, yet another was being born. Bravely defying eternity, life had no fear of creating one more. It continued its work and propagated according to its laws, indifferent to all hypotheses, forever toiling along its infinite path.*[61]

*Her thoughts then wandered to the deep gratitude she felt towards Pascal for having made her what she was. When, long ago when she was still a small child, he had taken her away from her horrible surroundings, he had surely acted out of the goodness of his heart. No doubt, he had also been eager to test how she would grow in an environment of truth and tenderness. It was something he always had in mind, an old theory he would have wanted to try out on a large scale: how the environment can provide culture, even cure, how it can improve and save the human being, both in the physical and moral sense.*[62]

These lines are tinged with an experimental Lamarckism found in many other passages by Zola, including in *Germinal. Germinal* can be taken as a parable in which an outsider helps regenerate a group. It is a kind of plea for more human compassion, altruism, care for the ill and infirm, and the social struggle against biological inequality.

This insight is not lacking in Charles Darwin's writings, in particular in *The Descent of Man*, published in 1871, but for Zola and his French contemporaries Darwin was first and foremost the man of the biological struggle, the emblem of the protagonists of *Money*.

**NOTES**

1. *Book of Proverbs*, Chapter 30, Verses 18-19.
2. Duret C., 1605, *L'Histoire Admirable des Plantes et Herbes Esmerveillables et Miraculeuses en Nature...*, Buon, Paris.
3. The use of this adjective may seem an anachronism. We shall nevertheless use it to denote either the emergence, generation, growth or transformation of living things as understood by the "Ancients". As soon as chronology permits, its use will be confined to the ideas at the origin of the movement spurred by Darwin's work.
4. This view of the origin of mankind explained Anaxagoras' refusal to eat fish which he took to be his ancestor.
5. We also encounter this idea in James Lovelock's Gaia hypothesis (see, for example, Lovelock J. E., 1988, *The Ages of Gaia. A Biography of our Living Earth,* Oxford University Press) which considers the Earth and biosphere as a superorganism (called Gaia). Even if the analogy may be of heuristic value, neo-Darwinians are hostile to the idea. To a neo-Darwinian, an organism is the product of natural selection which acts repeatedly on heritable variations. The Earth, however, does not multiply and give rise to several exemplars on which selection might act. Because selection occurs at the organ level, the organism contains organs whose prime function is to make it function! On the contrary, Gaia can only be a collection of items, each selected for its own benefit and for that of the species to which it belongs.
6. Plato, 360 BCE, *Timea*, 39e-40a and 69b-c. Translated by Benjamin Jowett. Internet Classics.
7. Aristotle, 30 BCE, *On the Soul,* II, 3, 415a-b. Translated by J.A. Smith. Internet Classics.
8. Arnould, J., 1996, *Les Créationnistes*, Bref n° 52, Cerf/Fides, Paris.
9. He also composed a poem, in grand baroque style, *Bacchus in Tuscany* (translated into English by Leigh Hunt in 1825): *"Dearest, if one's vital tide/ Ran not with the grape's beside/ What would life be (short of Cupid)? Much too short, and far too stupid."*
10. Quoted by Taylor, G.R., 1963, *The Science of Life. A Picture History of Biology*. Thames & Hudson, London, p. 111.
11. Quoted by Taylor, G.R, *ibid.*, p. 113.
12. Linnaeus, C., 1737, *Critica Botanica* (English translation by Sir Arthur Hort, published by Ray Society in 1938, Bernard Quaritch Ltd, London).
13. Linnaeus, C., 1787 (originally published in 1744), *Dissertatio botanica de Peloria.* In: Amoenitates Academicae, 3rd edn, Vol. 1, J.J. Plam, Erlangae, p. 68.
14. Linnaeus, C., 1789 (originally published in 1762), *Fundamentum Fructificationis.* In: Amoenitates Academicae, 2nd edn, Vol. 6, J.J. Plam, Erlangae, p. 296.

15. Marchant, J., 1719, *Observations sur la Nature des Plantes*, Mémoire de l'Académie Royale des Sciences, quoted by Guyénot, E., 1957, *Les Sciences de la Vie aux XVIIè et XVIIIè siècles. L'Idée d'Evolution.* Series: Bibliothèque de Synthèse Historique. L'Evolution de l'Humanité, Albin Michel, Paris, p. 375.
16. Duchesne, A.N., 1766, *Histoire Naturelle des Fraisiers.*
17. Quoted by Guyénot, E., *ibid.*, p. 377.
18. See insert below: "The doctrine of preformation".
19. The word fossil began to be used to denote something, or someone, old-fashioned only from the beginning of the 19th century onwards.
20. Quoted by Guyénot, E., *ibid.*, p. 340.
21. Quoted by Guyénot, E., *ibid.*, p. 341.
22. de Maillet, B., 1984, *Telliamed ou Entretiens d'un Philosophe Indien avec un Missionnaire Français sur la Diminution de la Mer.* Series: Corpus des Oeuvres de Philosophie en Langue Française, Fayard, Paris, pp. 248 and 304.
23. Voltaire, 1829, *L'Homme aux Quarante Ecus,* VI, in: Oeuvres, Vol. 34, Lefèvre, Paris, p. 43. Elsewhere, he wrote: "*Despite the great passion we have nowadays for family trees, there are few people who believe they descend from a turbot or cod.*"
24. Marquis d'Argenson, 1749, quoted by Guyénot, E., *ibid.*, p. 356.
25. Quoted by Taylor, G.R., *The Science of Life*, p. 78.
26. de Séchelles, H., *Voyage à Montbard*, quoted by Guyénot, E., *ibid.*, p. 357.
27. Quoted by Roger, J., 1989, *Buffon. Un Philosophe au Jardin du Roi,* Fayard, Paris, p. 537.
28. Roger, J., *ibid.*, pp. 540-541.
29. Rostand, J., 1966, *Un Grand Biologiste. Charles Bonnet, Expérimentateur et Théoricien,* Palais de la Découverte, Paris.
30. Quoted by Fischer, J. L., 1991, *La Naissance de la Vie. Une Anthologie*, Series: Agora. Les Classiques, Press Pocket, Paris, p. 15.
31. Quoted by Guyénot, E., *ibid.*, p. 385.
32. de Maupertuis, P., 1756, *Essai de Cosmologie*, quoted by Guyénot, E., *ibid.*, p. 391. The reference to "*elementary parts [that] had not kept the order they had in the father and mother animals*" might presage the notion of homology introduced later by Cuvier and Geoffroy Saint-Hilaire.
33. Diderot, 1971 (originally published in 1753), *De l'Interprétation de la Nature*, Editions sociales n° XI and XII, Paris, pp. 45-46.
34. Diderot, 1962 (originally published in 1830), *Le Rêve de d'Alembert*, Editions sociales, Paris, pp. 42-43. See *De l'Interprétation de la Nature*, n° L.
35. Herder, J.G., *Ideen zur Philosophie der Geschichte der Menschheit*, cited from Altner, G., 1983, Le tournant évolutionniste, *Concilium* 186, 46.
36. Voltaire, 1830, *Histoire du Docteur Akakia*, in: Oeuvres, Vol. 39, Lefèvre, Paris, pp. 483-484. On the subject of attraction, Voltaire went on to say: "*We let God create men in the way He sees fit without ever interfering; and each person is free not to believe that, in the uterus, the right toe attracts the left toe and the hand places itself at the end of the arm by attraction*" (*ibid.*, p. 497). His reference to snails is probably a reference to the comparison Maupertuis made in *Vénus Physique* between the mating behaviour of snails and the sado-masochistic practices that "*people who were cold from age, or whose senses were blunted, sometimes used [...] to arouse desire*".
37. Devillers, C., 1996, Homologie/Analogie, in: *Dictionnaire du Darwinisme et de l'Evolution*, Presses Universitaires de France, Paris.
38. Le Guyader, H., 1996, Geoffroy Saint-Hilaire/Cuvier, in: *Dictionnaire du Darwinisme et de l'Evolution*, Presses Universitaires de France, Paris, p. 1879.
39. Pichot, A., 1993, *Histoire de la Notion de Vie*, Series: Tel, Gallimard, Paris, p. 579.
40. Lamarck, J.-B., 1994 (originally published in 1809), *Philosophie Zoologique ou Exposition des considérations relatives à l'histoire naturelle des animaux; à la diversité*

*de leur organisation et des facultés qu'ils en obtiennent; aux causes physiques qui maintiennent en eux la vie et donnent lieu aux mouvements qu'ils exécutent; enfin, à celles qui produisent, les unes le sentiment, et les autres l'intelligence de ceux qui en sont doués*, Flammarion, Paris, I, 1, pp. 82-83.

41. Lamarck, *ibid.*, I, 1, p. 86.
42. Cladism is most advanced in this respect because it takes a monophyletic set of species (or individuals) only as a unit of systematics.
43. Lamarck, *ibid.*, I, 1, p. 80.
44. Subheading of the third chapter of the first part of the *Philosophie Zoologique*.
45. Lamarck, *ibid.*, I, 3, pp. 107-108.
46. Lamarck, *ibid.*, Chapter I, p. 8 (table of contents): *De l'Ordre naturel des Animaux et de la disposition qu'il faut donner à leur distribution générale pour la rendre conforme à l'ordre même de la nature.* (Of the natural order of animals and of the way in which they should be classified in order that this classification be in line with the order of nature).
47. Lamarck, *ibid.*, Chapter I, 3, p. 116: *Des espèces dites perdues.*
48. Pichot, A., *ibid.*, p. 654.
49. Lamarck, *ibid.*, I, 3, p. 117.
50. Into four types: peripheral or radial (Infusoria, Medusa), longitudinal (segmented animals), massive or mollusc-like (Molluscs, Rotifers ...), vertebrate.
51. Lamarck, 1815-1822, *Histoire Naturelle des Animaux sans Vertèbres.*
52. Lamarck refers to the motion of fluids and to a kind of autocatalytic increase in this motion both in the individual and in successive generations.
53. Lamarck, *ibid.*, I, 7, pp. 216-217.
54. The following passages which, doubtless, did little to enhance the prestige of transformism will illustrate what Lamarck really meant by "change through use": "*More than any other vertebrates, [birds] live up in the air in which they continually rise and which they traverse in all directions; the habit they have of filling their lungs with air to increase volume and become lighter has meant that this organ adheres to the sides of the breast, and that the air within, which is rarefied by the surrounding heat, can pierce the lung and its envelopes and penetrate nearly all body parts, from the big bones which are hollow to the quills of big feathers.*" "*Indeed, in birds, the air that penetrates the roots of hairs changes the base into a quill and forces these hairs to divide into feathers.*" (*Philosophie Zoologique*, pp. 148-149). "*When they are angry as is frequently the case in males, their strained inner feeling sends fluids preferentially toward this part of the head; horny matter is secreted in some, bony matter mixed with horny matter in others; these give rise to solid protuberances: This is how the horns and antlers, which arm the heads of most of these animals, came to be.*" (*Philosophie Zoologique*, p. 256)
55. For Pietro Omodeo (see article on Erasmus Darwin, in: *Dictionnaire du Darwinisme et de l'Evolution*, Presses Universitaires de France, Paris, 1996), "*the fact that both authors* [Lamarck and Erasmus Darwin] *had unusually well-matched ideas is doubtless due to a similar cultural background and to coinciding personality traits. Both had a wide range of interests ranging from meteorology to physics and chemistry, and were wont to ask themselves many questions on all kinds of subjects.*"
56. Pichot, A., *ibid.*, p. 680.
57. "*From deformed parents come deformed children, lame from lame and blind from blind, and, speaking generally, children often inherit anything that is peculiar in their parents and are born with similar marks, such as pimples or scars. [...] Such cases, however, are few; for the children of cripples are mostly sound, and there is no hard and fast rule regarding them.*" (Aristotle, 350 BCE, *The History of Animals*, VII, 585, translated by D'Arcy Wentworth Thompson.)
58. Lamarck, *ibid.*, I, 8, pp. 298-300.

59. Zola, E., 1969 (originally published in 1871) *Préface des Rougon-Macquart. Histoire Naturelle et Sociale d'une Famille sous le Second Empire*, Series: L'Intégrale, Vol. I, Seuil, Paris, p. 57.
60. Zola, E., 1970 (originally published in 1893) *Le Docteur Pascal*, Series: L'Intégrale, Vol. VI, Seuil, Paris, p. 518.
61. Zola, E., *ibid.*, pp. 654-655.
62. Zola, E., *ibid.*, p. 650.

# Chapter 2

# THE DARWINIAN PARADIGM

## 1. THE DARWINIAN REVOLUTION

Is there a Darwinian myth? When Karl Popper criticised Darwin's theory on the grounds that it could not be put to the test of non-falsifiability, he backed the idea that, like Marxism or psychoanalysis, Darwinism is a modern myth devoid of scientific rationality. Michael Denton, an Australian molecular biologist, adds:

> *Ultimately the Darwinian theory of evolution is no more nor less than the great cosmogenic myths of the twentieth century. Like the Genesis based cosmology which it replaced, and like the creation myths of ancient man, it satisfies the same deep psychological need for an all embracing explanation for the origin of the world which has motivated all the cosmogenic myth makers of the past, from the shamans of primitive peoples to the ideologues of the medieval church.*[1]

Denton uses the word myth in a reductive, even disparaging, way. To him, the theory of evolution, which claims to have changed our view of the world, is mere conjecture and conceals unavowed motives; it is just a fable. Although Neo-Darwinians have answered the criticisms raised in his book *Evolution: a Theory in Crisis*, Denton still maintains that:

> *the "mystery of mysteries" - the origin of new beings on earth - is still largely as enigmatic as when Darwin set sail on the Beagle.*[2]

He continues to challenge the idea that the Darwinian paradigm has any scientific substance. In our view, his attitude is rather extreme.

**Karl Popper and Thomas Kuhn**

In *Logic of Scientific Discovery* (1959), the Viennese philosopher Karl Popper put forward the epistemological criterion of "refutability" (or "falsifiability") whereby a theory is said to be scientific if, from its general terms, we can draw a term that describes at least one empirical test that may refute it. For quite some time Popper criticised the classical theory of natural selection (formulated in 1944, in *The Poverty of Historicism*). The two propositions to which this theory is often reduced (the one who has survived is the fittest, the fittest is the one who has survived) are, he said, circular arguments constituting a simple irrefutable case of tautology. In 1972, in *Objective Knowledge*, taking a less adamant view, he conceded that:

*Darwin's theory of natural selection showed that it is in principle possible to reduce teleology to causation by explaining, in purely physical terms, the existence of design and purpose in the world.*[3]

He concluded:

*Darwin showed that we are all completely free to use teleological explanation in biology - even those of us who happen to believe that all explanation ought to be causal.*

Why did Popper make such an about-turn? He would explain later that, in the meantime, he had learnt about the synthetic theory of evolution with which he agreed. However, Dominique Lecourt notes that:

*this rallying cry was expressed in words that did not present the theory as a "scientific" theory according to the meaning Popper gave to the word "science", but as a doctrine with the characteristics of "metaphysical research programmes", words that parodied (and criticised) those of his dissident disciple Imre Lakatos.*[4]

The epistemologist Thomas Kuhn denoted *"paradigms"* those scientific achievements with the two following features:

*Their achievement was sufficiently unprecedented to attract an enduring group of adherents away from competing modes of scientific activity. Simultaneously, it was sufficiently open-ended to leave all sorts of problems for the redefined group of practitioners to resolve.*[5]

The examples he gave were the *Physics* of Aristotle, *Almagest* of Ptolemeus, *Principia* and *Optiks* of Newton, *Electricity* of Franklin, *Chemistry* of Lavoisier and *Geology* of Lyell.

## 1835: A story about finches

In 1835, an English ship, The Beagle, made a stop-over at the Galapagos Islands, an archipelago in the Pacific Ocean off the coast of South America. During the halt, a young naturalist, Charles Darwin (1809-1882) collected many specimens of birds, mainly mocking-birds and finches and, on his return to England, in October 1836, gave them to the ornithologist John Gould to be examined. Gould identified three distinct species of mocking-birds and thirteen species of finches. Why did Darwin seek his expert opinion?

The similarities and differences among the birds had intrigued Darwin. The feathers and colours of the mocking-birds were so similar that, at first, he thought the birds belonged to the same species. Was he to conclude that the distinction between species and varieties was somewhat arbitrary? What would then become of the dogma of the fixity of species?

The finches (later to be known as the famous "Darwin finches") differed chiefly in the shape of their beak and in their habits. Some lived at ground level and ate mostly seeds; others lived in trees and fed off insects. There were also differences in their geographic distribution. Vice-Governor Lawson told Darwin that they came from different islands and that the natives had no difficulty in telling them apart. Darwin admitted:

> *It never occurred to me that the production of islands only a few miles apart, and places under the same physical conditions, would be dissimilar.*[6]

Darwin added an extra piece of information to the observations he made on the Galapagos Islands. "His" finches were very similar to those in South America. He therefore speculated that a few birds of a single species had migrated to the archipelago where they had then diversified. His island observations thus supported his belief in transformism and his intuition that one species can give rise to another. As we have seen, this idea was not really novel. Lamarck and Charles' grandfather, Erasmus Darwin, had already mentioned it at the beginning of the 19th century.[7]

Just like Lamarck, Darwin noted how well the structure and performance of the organs of living beings match their way of life; how well they are adapted to it. In *The Origin of Species*, he even acknowledged:

> *It is, however, difficult to decide, and immaterial for us, whether habits generally change first and structure afterwards; or whether slight modifications of structure lead to changed habits; both probably often occurring almost simultaneously.*[8]

He gave the following example that could well have come from Lamarck's work:

*In North America the black bear was seen by Hearne swimming for hours with widely open mouth, thus catching, almost like a whale, insects in the water.*[8]

Unlike Lamarck, however, Darwin explained the gradation between so many species, varieties, and even individuals by natural selection. This principle will become the mainstay of the theories belonging to the Darwinian paradigm, defined by Jean Gayon as follows:

*Any view that considers evolution as a gradual change in species which arises chiefly by a process of natural selection acting upon a variation within populations is "Darwinian".*[9]

## The Darwinian theory of the origin of species

We tend to forget that Darwin did not mention "evolution" in *The Origin of Species*. The term was already in use but by contemporary embryologists who were not that interested in the origin of species. It referred to the doctrine of preformation (see Chapter 1, how the word evolution was introduced by Charles Bonnet) and had the opposite meaning to the one Darwin was to give it.

*There is nothing more remote from the idea that new species pre-exist in old species, from which they would just need to evolve (=unfold) with time, than Darwin's doctrine.*[10]

Like Lamarck, Darwin spoke initially of transmutation, then of transformation of species. He innovated, however, by inverting the traditional relationship between individual and species. For Darwin, species were not types to which individuals had to conform as closely as possible if they were not to be considered monstrosities. On the contrary, he thought that individuals changed as a result of natural selection and ended up by modifying the species. The very notion of a species thus lost clarity:

*Certainly no clear line of demarcation has as yet been drawn between species and sub-species - that is, the forms which in the opinion of some naturalists come very near to, but do not quite arrive at, the rank of species: or, again, between sub-species and well-marked varieties, or between lesser varieties and individual differences. These differences blend into each other by an insensible series; and a series impresses the mind with the idea of an actual passage.*[11]

Darwinian theory is based on the following facts and principles:

There is a considerable and apparently inexhaustible variability within species due to often very slight individual differences. This variability, which Darwin noted but could not explain, is either discontinuous or continuous. (Darwin took continuous variability to be the more important of the two for evolution). Darwin was convinced that:

> *the most experienced naturalist would be surprised at the number of cases of variability, even in important parts of structure [...]. It should be remembered that systematists are far from being pleased at finding variability in important characters ...*[12]

Darwin's view of biodiversity clashed with Linnaeus' ideas because the emphasis was no longer on the specificity of variations between individuals but within individuals.

If the emergence of living beings follows a geometric progression, then there are more creatures than natural resources can support. Because of competition, only a few will survive. Darwin was forced to introduce the idea of a *struggle for life* inspired by Malthus' thinking:

> *Hence, as more individuals are produced than can possibly survive, there must in every case be a struggle for existence, either one individual with another of the same species, or with the individuals of distinct species, or with the physical conditions of life. It is the doctrine of Malthus applied with manifold force to the whole animal and vegetable kingdoms; for in this case there can be no artificial increase of food, and no prudential restraint from marriage.*[13]

Survivors display favourable individual differences. Nature, like breeders, sorts out living beings and selects on the basis of their natural variability. Darwin wrote:

> *This preservation of favourable individual differences and variations, and the destruction of those which are injurious, I have called Natural Selection, or the Survival of the Fittest. [...] it implies only the preservation of such variations as arise and are beneficial to the being under its conditions of life.*[14]

Selected organisms pass their advantage on to their descendants, thus modifying the original "type" slightly. When this process is repeated over thousands, even millions, of generations, it creates new species.

> *Thus the small differences distinguishing varieties of the same species, steadily tend to increase, till they equal the greater differences between species of the same genus, or even of distinct genera.*[14]

The idea of natural selection was brewing towards the middle of the 19th century even though it had not yet been given a name. This is clear from a letter Darwin received in 1858 from a young colleague, Alfred Russel Wallace (1823-1913). Wallace was a professional collector who had spent several years in the Brazilian forests. He was a self-taught naturalist and had published in 1855 an article *On the law which has regulated the introduction of new species.* In it, he had tried to solve a problem left unanswered in Charles Lyell's *Principles of Geology* (1832). Under what conditions does a new species emerge? Wallace stated that there were wide variations within a species and noted the importance of bio-geographic conditions. He wrote as follows to Darwin from an island not far off from the Spice Islands:

> *There is no limit of variability to a species, as formerly supposed. Useful variations will tend to increase; useless or hurtful variations to diminish...*[15]

However, Wallace could not answer Lyell's question. He observed affinities among successive species and thought that each species is born in geographic and geological coincidence with a related, pre-existing species but did not know how. After reading the letter, Darwin wrote to his friend Lyell:

> *Wallace [...] has today sent me the enclosed [...] Your words have come true with a vengeance - that I should be forestalled [...] I never saw a more striking coincidence; if Wallace had my MS sketch written in 1842, he could not have made a better short abstract!*[16]

This episode in Darwin's life has fascinated historians of science and psychologists alike. Why did Darwin take such a long time to publish his conclusions? Did he have scientific reservations or did he fear conflict with an establishment ready for a fight? What should we make of the terms on which he was with Wallace and of the advice offered by his entourage in this quarrel about priority of claim and scientific paternity? We shall not delve into the details but just recall the main historic facts: Two notes (that of Wallace and of Darwin) were read at the Linnean Society of London on the 1st July 1858; *The Origin of Species* was written at lightning speed and finished on the 1st October 1859; the 1250 copies of the first edition were sold on the very day they were published, the 24th November of the same year. This series of events was followed by rebuttals, debates and verbal jousts in which "Darwin's bulldog", Thomas Henry Huxley (1825-1895), was a high-profile figure.

Two points need to be emphasised at this stage.

1. There were differences in Wallace's and Darwin's hypotheses on the modification of natural species. As Jean Gayon points out in the first chapter

of his study *Darwin and after Darwin*, although both hypotheses were put forward in 1858, they were hardly equivalent. Wallace stressed competition among varieties (or races). Advantaged varieties expand at the expense of other varieties which they end up by eliminating. Darwin did not deny this but explained the modification of species by the accumulation of individual inherited differences.

Besides, Wallace did not accept a model inspired by domestic breeding. He criticised both Darwin's method and concept. Darwin, on the other hand, maintained that data on domestic animals were a relevant foundation for his hypothesis.

> *The detour via domestic breeding was not an artifice to put across his ideas. It was essential from the point of view of method. Without it, the subtle collusion between variation, heredity and change that is a feature of Darwin's selection hypothesis would have been mere speculation devoid of empirical content.*[17]

To defend his hypothesis, Darwin unhesitatingly took advantage of all observations gathered over more than 20 years in five fields: palaeontology (fossils, the historical archives of life), bio-geography (the distribution of species over the surface of the Earth, explained by descent accompanied by change), systematics (a hierarchical rank-order of species produced by descent, and not by a divine plan or by man's imagination), morphology (the existence of homologies in shape in living things) and embryology (embryogenesis which holds the traces of the sequence in which species have evolved).

2. How to interpret natural selection has posed, and still poses, problems. Darwin was already aware of this when he wrote:

> *It has been said that I speak of natural selection as an active power or Deity; but who objects to an author speaking of the attraction of gravity as ruling the movements of the planets? Every one knows what is meant and is implied by such metaphorical expressions; and they are almost necessary for brevity. So again it is difficult to avoid personifying the word Nature; but I mean by Nature, only the aggregate action and product of many natural laws, and by laws the sequence of events as ascertained by us. With a little familiarity such superficial objections will be forgotten.*[18]

## The most serious objections to Darwin's theses

When *The Origin of Species* was published, Lord Kelvin (William Thomson, 1824-1907) was one of the greatest scientists in Britain. A profoundly religious man, he attacked the chronological aspects of Darwin's theory. If we are to believe Darwin, he said, the geological deposits of the Tertiary era date back 300 million years but if we estimate the total life-span of our star from the rate at which the sun's energy is dissipated, it is shorter than this, just a few million years. (Kelvin mistakenly took the sun to be an incandescent molten mass, neglecting thermonuclear reactions which were unknown at the time.) Since the sun's age could not be greater than its lifespan, Kelvin concluded that our star, and therefore the Earth, was young.

Kelvin also attacked the work of geologists, especially in *The Doctrine of Uniformity in Geology Briefly Refuted* (1865). According to uniformitarian theory which was latent in Buffon's work before being developed by Lyell, the earth's crust was formed by gradual natural processes and not by cataclysmic events. Darwin, by considering that descendants are modified through variations and natural selection, was in fact applying this geologic concept to life. He was thus directly concerned by Kelvin's criticism of uniformitarianism. He took it for a strong blow and in 1871 wrote to Wallace:

> *I have not as yet been able to digest the fundamental notion of the shortened age of the sun and earth.*[18]

He suggested that evolution might have occurred over a shorter time-span and, risking self-contradiction, that higher-order organisms might evolve faster than lower-order organisms. Taken to task, Wallace suggested that originally the earth's orbit might have been less eccentric and thus induced changes in some parameters and biological rhythms. Whatever the explanation, the "odious spectre" of Kelvin's thesis was to haunt Darwin until his dying day.

The Scot Fleeming Jenkin (1833-1885) made an even more serious objection. In 1867, in an article in the *North British Review* (which was more like a satire than a genuine scientific article), he asked Darwin two fundamental questions:

1. Is hereditary variation continuous or discontinuous? Is continuous variability inherited or are there isolated variations?

2. What is the mode of hereditary transmission? Does the new trait predominate (this would ensure the success of the selective process) or, as current opinion had it and as Darwin himself believed, were the children intermediate between both parents? If so, interbreeding would eliminate the new trait immediately. For this not to happen, the trait would have to emerge in many - even most - individuals at once. But what then becomes of the hypothesis of random variations on which the theory of natural selection is based? Does this not imply an internal, non-random variation force?

Darwin admitted that Jenkin's questions were the most apt he had ever been asked. Yet again, he will try to answer, and fail. Jenkin's questions will in fact only be answered with the development of biometrics and population genetics. Highlighted by J. Gayon, this Darwinian dilemma reveals a major ambiguity in his work:

> *Darwinian natural selection is, first of all, a principle whereby variations that have appeared in an isolated manner in individuals gradually accumulate [...]. But Darwin also presents natural selection as acting on a variation that is minus a bit or plus a bit, in other words, on a continuous variation. In this instance, the variation is widespread, and selection must then be seen as a model for shifting the average character.*[19]

To escape from this dilemma, Darwin put forward a hypothesis called pangenesis. Reproductive elements do not engender new individuals, but all cells - the units of the whole body - do. They emit gemmules that aggregate in the reproductive elements so that, when the time is ripe, they can form cells similar to the original cells. Thus,

> *if the varying individual did not actually transmit to its offspring its newly-acquired character, it would undoubtedly transmit to them, as long as the existing conditions remained the same, a still stronger tendency to vary in the same manner.*[20] Had Darwin forsaken his hypothesis of natural selection acting on random variations? He ended up by admitting: *There must be some efficient cause for each slight individual difference, as well as for more strongly marked variations which occasionally arise; and if the unknown cause were to act persistently, it is almost certain that all the individuals of the species would be similarly modified.*[21]

In fact, neither Darwin nor Jenkin had the means to answer this question. There were thus two escape routes: either to consider that selection acts according to a continuous spectrum, i.e. a Gaussian distribution - this view sprung directly from Darwin's work - or to forget heredity by blending and apply natural selection to discrete entities instead. This will be the path that a monk who loved peas will follow ... but beforehand we must mention a great 19th century theoretician of biology, August Weisman.

## Was August Weisman the father of neo-Darwinism?

It has become the custom to call the synthetic theory of evolution proposed by Theodosius Dobzhansky, Ernst Mayr, George Simpson and others in the 1930-40s "neo-Darwinism". According to George Simpson, the explicit reference to Darwin is due to the importance that the synthetic theory confers to natural selection; the prefix "neo" distinguishes the theory from Darwin's own theory in that the neo-Darwinian paradigm takes Mendel's laws into account.

However, a claim had already been lain on the expression neo-Darwinism in the 19th century by the German biologist August F.L. Weisman (1834-1914). His claim was typical of the squabbles of the time when biological circles fought over who was Darwinian and who was not. Weisman, who uncompromisingly defended selectionism and refuted all other mechanisms, considered himself to be Darwin's sole successor. So did his contemporary, G.J. Romanes, who placed the theories of Lamarck and natural selection on a par. We shall not take sides but focus on how Weisman became a fierce adversary of the theory of acquired inheritance and came to be seen as the person who refuted it.

By background and training Weisman belonged to a group of researchers who believed in reductionism; disciplines such as cytology or embryology required knowledge of physical chemistry. After practising medicine and presenting a thesis on the metamorphosis of insects, Weisman announced his adherence to Darwinism during his inaugural lesson at the University of Freiburg-in-Bresgau (*Über die Berechtigung der Darwin'schen Theorie*). However, in his early work he had betrayed no qualms about the inheritance of acquired characters. It is only between 1881 and 1883 when studying maximum life-span as a specific trait subject to natural selection that he convinced himself that there was no way of transmitting the results of training to future generations.[22] How could "life-span" be acquired and passed on directly? The hereditary determinant cause must precede the effect. At the same time, Weisman introduced the idea of an immortal cell line, the germ cells (*germen*), where a succession of cells leads to the gametes. It is this cell line that transmits traits from one generation to the next "through" somatic lines (*soma* = all the non-sexual body cells).

Gustav Jaeger (1878) and Moritz Nussbaum (1880) had already had the idea of a continuous germ line but Weisman was the first to use it as a basis for an evolutionary theory of heredity. He asked himself what was the source of the variations on which natural selection acts and thought it might be sexual reproduction. In 1886, he gave a conference entitled: "*Sexual reproduction and its relevance to the theory of natural selection*" ("*Die Bedeutung der sexuellen Fortpflanzung für die Selektions-Theorie*".)

In 1887, in the wake of Oscar Hertwig's work on fertilisation, Weisman suggested that germinative plasma might be found in the filaments of dividing cells. Hertwig was professor of anatomy at Berlin and the first, in 1875, to observe the fusion of two nuclei in a fertilised egg. Hertwig formulated the general idea that fertilisation consists mainly in the fusion of two cell nuclei, one of maternal origin (that of the ovule) and the other of paternal origin (that of the spermatozoa). At the end of the 1870s, Eduard Strasburger described the clonal division (mitosis) of plant cells and noted the presence of easily stained particles in the daughter cells. Walter Flemming (who extended these observations to the animal kingdom) called their substance chromatin before Wilhelm Waldeyer, in 1888, finally called them **chromosomes**. Pierre-Joseph van Beneden established that the nuclei of ovules and spermatozoa contain the same number of chromosomes whereas the nuclei of the cells that give birth to them contain twice as many. In 1887 Theodor Boveri confirmed this discovery and generalised it as the law of chromatic reduction. When sex cells divide (meiosis), they only receive a single set of parent chromosomes; a double set is formed in the fertilised egg. At the time Weisman wrote:

> *The brilliant research on the process of fertilisation which was initiated by Auerbach and Bütschli and pursued by Hertwig, Fol and Strasburger right up until the work of van Beneden - not to mention many other distinguished names that belong to this list - as well as the theoretical considerations put forward by Pflüger, Nägeli and myself lead to at least one firm result. This result is the existence of a hereditary substance, namely, a physical vehicle for hereditary trends, which is contained in the nucleus of the germ cell, in that part of the nuclear filament which at times has the shape of handles or short rods (chromosomes).*[23]

Weisman went on fighting the idea of the inheritance of acquired characters. In 1988, for example, he criticised "*the would-be botanical proof for inheritance of acquired characters*". Four years later, he proposed a modified and much more complex theory of the germinal line whereby the germinative plasma is composed of "biophores" which are structures that can grow and reproduce by division. All biophores taken together determine the characteristics of a cell. Weisman spoke of a "determinant". In multicellular creatures, several determinants are needed and are combined into an "ide". Several ides make up an "idante" which, according to Weisman, was equivalent to a chromosome.

Weisman's contribution to modern biology and to the theories of evolution was far from negligible. He combated the idea of the inheritance of acquired characters and introduced a fundamental distinction between *soma* and *germen*.

## 2. THE ORIGINS OF GENETICS: A STORY ABOUT PEAS

Against this backdrop, biologists turned toward acquiring an understanding of the mechanisms of inheritance. Gregor Mendel's unpretentious experiments in botany were to provide a first insight.

The 18th century had witnessed a series of experiments on the hybridisation of plants by Joseph Gottlieb Kölreuter, William Herbert, Christian Konrad Sprengel, Thomas Andrew Knight, and others. These were pursued into the 19th century. In France, in 1825, Augustin Sageret made crosses of the so-called "*cantaloup*" and "*chaté*" varieties of melons, which have different grooves and netting. He studied the combination of these two characters in hybrids. Apparently, at about the same time (probably before 1829), the Swiss pharmacist J.A. Colladon recorded observations on the interbreeding of grey and white mice. However, all this research was not really to take off until two botanists, Naudin and Mendel, published their studies in 1865 although Mendel's results were to remain poorly understood until the end of the 19th century.

Charles Naudin (1815-1899), who worked at the Botanical Gardens (*Jardin des Plantes*) in Paris, was interested in defining species. His early opinion on the matter belonged to a kind of limited transformism. In "*Philosophical Considerations on Species and Variety*" ("*Considérations Philosophiques sur l'Espèce et la Variété*"), he wrote:

> *We do not believe that, when nature created species, it acted in a way any different from the way we act to create varieties; in plain words, we have transposed nature's process into our practice.* What is this process? It is to attune *each member with the whole by assigning to it the function it must carry out in the general organism of nature, a function that is its raison d'être.*

He who selects decides which individuals deviate from a specific type, weeds them out, and, after many generations, obtains a fixed variety, that is an artificial species. From 1856 onwards, Naudin carried out interspecific hybridisations and tried to obtain new stable forms but in vain. He then put forward several hypotheses later to be formulated with greater precision by Mendel. Among them were the uniformity of first-generation hybrids (already noted by Cotton Mather in 1716), the identity of reciprocal crosses, the return to parental types and the heterogeneity of generations from the second cross onwards. Finally, he postulated that the two specific essences in pollen and in the ovules of hybrids are separate. William Bateson will call this the *purity of the gametes*; pollen and ovule are either of the paternal or maternal type. In spite of all these advances in knowledge, the mechanism of inheritance will not be formulated for the first time in the Botanical Gardens of Paris but in the experimental vegetable patch of a Moravian monastery.

If things had gone as expected, the young Johann Mendel (1822-1884) should have become a peasant like his father. Bound by duty to the lord of the manor, he would have cultivated the chalky soil of Northern Moravia. However, the schoolteacher and the priest of his native village convinced his parents to let him pursue his studies and so, at the age of 21, Mendel entered the monastery of Saint-Thomas of the Augustines of Brünn (now Brno). He was given the name of Gregor and began to study theology but also physics,[24] zoology, botany and palaeontology. He went to the University of Vienna from 1851 to 1853 and was back in Brünn in 1856 but without having obtained his civil service teaching diploma. Until his election as head of the monastery in 1868, he taught at the town's state college and looked after an experimental garden created by an abbot of the monastery in 1830. It was in this 7 m by 35 m plot that he tended a "medley of forms". He cultivated 27 000 pea plants belonging to 34 varieties and examined 12 000 plants and their 300 000 seeds; a truly laborious undertaking.

In 1862, Mendel co-founded the Natural History Society of Brünn (the *Naturforschender Verein*) where on February 8 and March 8, 1865, he presented his results on the hybridisation of peas, later published in the Society's Transactions. The paper was sent to several European biologists in Berlin, Vienna, and London (it was sent to the *Royal Society* and the *Linnean Society*) as well as to biologists in America. It was a kind of treatise on the transmission of characters in true-breeding lines (which he called species) and in hybrids. Whereas the characters of true-breeding lines are preserved from one generation to the next, in hybrids, characters that were united in the first-generation are separate in the descendants; there is a return to the types of the original line. In stating his laws on variability in the descent of hybrids, Mendel was in fact making explicit the empirical rules that farmers had used for thousands of years to select species by successive crosses. He also postulated that transmitted characters are determined by factors present in the sex cells which ensure fertilisation.

It now seems certain that Darwin was informed of Mendel's work by the German botanist Heinrich Hoffmann.[25] However, it was not until 1900 that Mendel's results were reproduced by three botanists working independently. What was the reason for this lack of interest leading to oblivion? Was Mendel the unknown genius or accursed scholar some authors have portrayed? This portrayal is a caricature even if it does hold an element of truth. Mendel was certainly misunderstood by his colleagues; he did not seek to impress his results upon them at all costs. Maybe, his "mistake" was to have stepped out of the 19th century with a hypothesis-based deductive method and reductionist views without entering the 20th century since he was unaware of the existence of chromosomes.

Mendel's approach to experimental method is one of the novelties of his work. Instead of interpreting the results of crosses in an inductive manner as hybrid producers do, he saw cross-breeding as an experimental tool for confirming or refuting a predefined theoretical model. His work was probably trying to test the hypothesis of the fixity of species. Cross-breeding should induce no change in the type inherited from the parents. Mendel did not know about mutations and his genetics was thus incompatible with evolutionism. It is therefore hardly surprising that researchers like Darwin took no interest in his work.

Mendel cultivated two varieties of peas with few but marked differences: giant or dwarf peas, green or yellow, smooth or wrinkled. Each seed was sown separately as were the hybrids obtained by manual hybridisation. He used contemporary pre-statistical methods based on the law of large numbers but not sampling theory. He did not trace the individual lines of descent but calculated the ratios applicable to the transmitted characters. These ratios are correctly formulated in the limerick about a young woman who had an affair with a coloured man:

The results of her sins
Was quadruplets, not twins -
One black and one white and two khaki.[26]

Mendel wanted to break away from the vague vocabulary hybrid producers used to describe relationships among descendants of hybrid lines. In particular, he wanted to clarify the concept of inheritance in hybridisation.

Tradition has it that Mendel formulated several laws; we usually speak of three: "the law of the uniformity of the first generation", "the law of segregation" (or "the law of the purity of the gametes") and the "law of the independence in the segregation of characters" (or the "law of independent assortment"). It will be later discovered that there are exceptions to this last law. First, genes close to each other on a chromosome do not segregate independently; they are said to be linked. Second, because genes co-operate in the making of organs and organisms and in their function, their independence is relative. Moreover, the ratios expected from the second law are not always correct. There can be segregation distortion (*see Chapter 5*) or gene conversion of one of the alleles during meiosis in the hybrid, so that the two parent characters no longer have the same frequency in the offspring.

Mendel's three laws are closely linked with his 1865 paper entitled *"Research on Hybrid Plants" ("Versuche über Pflanzen-Hybriden").*

*However, the didactic way in which they are traditionally presented does not reflect Mendel's ambition. Mendel was not looking for laws but for a single law, specifically a general law on the formation and development of hybrids.*[27]

The notion of development is essential. Each hybrid with a given character has offspring that are distributed precisely. Mendel tried to predict the proportions in which this character was distributed over several generations using a mathematical law for series expansion.

**How Mendel crossed peas**

Mendel performed his experiments on autogamous varieties of the edible pea (species: *Pisum sativum*) that were well fixed (and that he therefore called species). He first sought to find true-breeding lines by methodically harvesting the seeds from each self-fertilised plant over several generations. He then crossed these pure lines by artificial pollination. In this way, he was able to pair varieties with specific differences (round x wrinkled; yellow albumen x green albumen; white flowers x coloured flowers; simply inflated pods x constricted pods; axile flowers x terminal flowers; short stems x long stems).

For each pair, he obtained uniform first-generation (F1) offspring that reproduced one of the two parental forms. He then let natural self-fertilisation work for the next few generations (F2, F3, F4). For the F2 generation, he repeatedly obtained a three-quarter of plants of the type present in F1 and one quarter of the type concealed in F1. It is here that he introduced the notion of dominance. He called the type that was present in F1 dominant (D) and the type that was concealed recessive (r).

Mendel then self-fertilised the F2 plants to obtain the F3 generation. A third of the dominant-type F2 plants produced exclusively D plants whereas the remaining two-thirds produced three-quarters of D plants and one quarter of r plants. The recessive-type F2 plants gave r plants only.

Mendel used gamete purity to explain his results. In a hybrid, for a given character, each reproductive cell retains and passes on to the next generation just one of the two characters that are united in the hybrid. This has an important conclusion:

*The product of the association of two gametes with a similar trait is totally devoid of the hybrid character of the parent. It is as if the individuals A that arise from the hybrid Aa were freed from the weight of their ancestry.*[28]

Mendel also made crosses involving other characters and concluded that each pair of characters is inherited according to the same laws, as if it were alone involved.

Whether Mendel formulated just one law or several does not really matter; the important point is that he made a decisive conceptual contribution to biology. He was the first to consider the inherited characters of an organism as entities that are transmitted independently of one another and that are influenced neither by hybridisation nor by the environment. In other words, living beings are a mosaic of characters each of which is inherited independently. Mendel's view thus clearly opposed the theory of inheritance by blending. This is probably the main reason why he was rejected by the scientific community of his time, in particular by Carl Nägeli.

*In a letter to Nägeli, Mendel stresses his hypothesis. At each generation, the two parent characters appear separate and unchanged, and there is nothing to indicate that the one has taken anything at all from the other.*[29]

Whereas Nägeli wrote in 1865:

*In general, the characters of the parent forms are passed on so that both influences are found in each hybrid individual. It is not true that a character can be passed on in an unchanged form by one of the parents, and another character in an unchanged form by the other parent. On the contrary, there is interpenetration of parental characters and a fusion of the two characters.*[29]

These two passages illustrate clearly Nägeli's disagreement with Mendel. He suggested that Mendel cultivate other plants such as *Hieracium*. Unfortunately, this plant, in which he was an expert, presents several disadvantages. Its flowers are so small that it is difficult to fertilise manually and the hybrid forms behave "aberrantly" because - and this was not known at the time - the plant can reproduce by parthenogenesis. Mendel was thus unable to reproduce his results on peas.

The Mendelian legend is the counterpoint to the Darwinian myth. Whether the reported facts are true or not, this legend signposts the entry into a new era in the history of biology, the era of genetics.

## 3. BIOMETRICIANS VERSUS GENETICISTS

Darwin's and Mendel's followers could not ignore each other for long. Their first clash occurred on a rather singular stage. The Darwinians had gathered under the banner of a new discipline, biometrics, and "*everything leads us to believe that the controversy between biometricians and Mendelians was essentially an English affair*".[30] This seemingly local quarrel is a highly interesting event in the history of evolutionary theory. What was the issue at stake?

The question that baffled the last two decades of the 19th century and, especially, the first two of the 20th century was whether the Mendelian model of discontinuous inheritance could be reconciled with the biometricians' Darwinian view of evolution? Many researchers believed these two approaches to life to be incompatible. To reconcile them would be a crucial step toward a unified vision of life. The task will befall to population genetics but, beforehand, let us examine the two conflicting views.

## Galton and Pearson: The founders of biometrics

Biometrics is the statistical study of variations in a group (or population) of individuals of the same species. A Belgian physicist and statistician, Adolphe Quetelet, was at its origin. Quetelet recorded the variations in height of a group of human subjects and depicted these variations as a frequency polygon (*Anthropométrie* (1871)). We still use this graphical representation when studying the statistics of characters present in homogeneous populations of many animals or plants. However, the true founder of biometrics was a young cousin of Charles Darwin, Francis Galton (1822-1911). Attracted by Darwin's theory and particularly interested in the transmission of variations, Galton developed a tool - mathematics - that had been little used in biology until then. Galton published *Typical Laws of Heredity* (1877), *Family Likeness in Stature* (1886) and *Natural Inheritance* (1889) and founded the English school of biometrics for the application of statistical methods to the study of living things. Although he did show some interest in the size of peas, he was particularly drawn to the study of the human species even though this was poor study material for biologists, experimental and directed interbreeding being out of the question. Notwithstanding, Galton began systematic research work on twins at the end of the 19th century.

In 1869, he published *Hereditary Genius: An Enquiry into its Laws and Consequences*. In this work, he used for the first time the bell-shaped normal distribution curve to describe what he called the "law of deviation from an average".[31] He chose these words because of the "trait" under study, namely, human genius. Is not a man a genius because of the deviation from the average rather than because of genius itself? Galton tried to use the bell-shaped curve to prove that the variations he measured were inherited but his measurements of human genius were rather imprecise and incongruous. In the mid-1870s, he therefore turned to the study of sweet peas and published his results in 1877 in a work - *Typical Laws of Heredity* - describing reversion. Reversion is the tendency offspring have to look like their forebears. In his investigation of the size of pea seeds, Galton observed that larger-than-average individuals, when crossed, have offspring of a size closer to the average. This publication was the first attempt to present natural selection in statistical terms.

In 1884, at the international exhibition on medicine at the Science Museum in South Kensington, Galton offered to give visitors their anthropometric measurements for the sum of three shillings; 9337 people accepted the offer. He used the funds to set up an anthropometry laboratory for the collection and treatment of such measurements. In 1885, he published *Regression towards Mediocrity in Hereditary Stature* from which Figure 2.1 is drawn. He showed yet again that the mean filial regression towards mediocrity is directly proportional to the parental deviation. From these results on human stature, he could generalise what he had initially called reversion and henceforth denoted regression.[32] (The word is still used in statistics; we refer to a regression line).

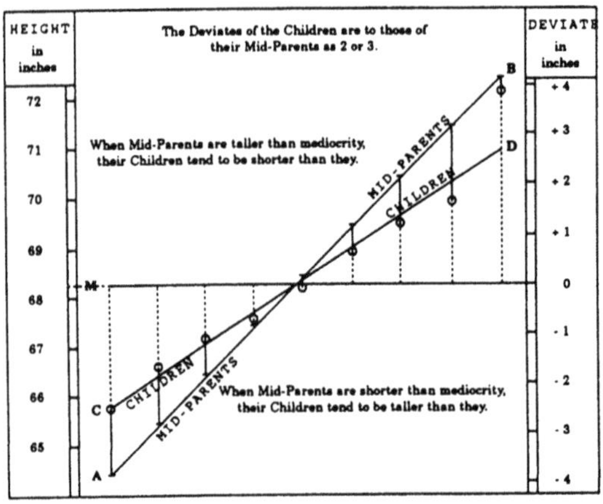


*Figure 2.1.* From Galton, F., *Regression towards Mediocrity in Hereditary Stature,* 1885.[32]

**A comment on regression**

Galton's mistake was that, from the second generation onwards, he no longer considered individuals (or, better still, an individual) but the mean of the offspring; this mean relayes to a set of individuals with very different characteristics. The problem will not be solved until about twenty years later by the work of Wilhelm Johannsen (see below).

A fundamental idea arising from Galton's work on sweet-peas was that the individual inherits not only from his parents, but also from his ancestors. Galton spoke of ancestral heredity and introduced the idea, contrary to the Darwinian view, that heredity is a force acting from a distance. Why should this idea upset the theory of natural selection? Because if we accept regression - a return to an average ancestral type - any gradual transformation of populations by selection and transmission of variations in continuous characters, from one generation to the next, becomes illusory. In other words, Galton's hypothesis implies that the mean of a population cannot be shifted without abrupt jumps from one stable state to another. (Nowadays, this is called saltationism).

Galton was indeed a strange man. He set the foundations for a new science - biometrics - but ended up promoting saltationism after having been an early supporter of the theory of natural selection. These were the two lines of thought that clashed violently in England at the beginning of the 20th century.

Biometrics may have not developed as it did if it had not received the support of Karl Pearson (1857-1936), a mathematician at University College London, who was invited to take an interest in biology by his colleague, the zoologist Walter Weldon (1860-1905). Pearson pursued Galton's work and applied mathematics to the study of variation and heredity. He devised several statistical methods of which the best known are the chi-square test - applicable to the overall quadratic deviation - and the coefficient of correlation. What were these tests for? The title of one of Pearson's articles tells us: "*On the criterion that a given system of deviations from the probable in the case of a correlated system of variables is such that it can be reasonably supposed to have arisen from random sampling*".[33] Pearson will attempt to derive several laws: "ancestral heredity" (a legacy from Galton) states that each individual is a combination of all his ancestors and "monotypia" that each of us is born with undifferentiated organs. All his work is in statistical terms.

Pearson was an eager advocate of natural selection and of gradual and continuous evolution, unlike Galton who was willing to accept that species could appear by a single variation or a sudden jump. If Galton is taken to be the "historical" founding-father of biometrics, then it might not be absurd to consider Pearson its "scientific" founder. It was Pearson who founded the journal *Biometrika* in 1901.

The philosophical and ideological undercurrents of their work prompt two questions. First, what were their links with the eugenics school of thought at the end of the 19th and 20th centuries (we shall come to this in the next part of this chapter); second, how were they in conflict with the partisans of Mendelism. Pearson's approach can be said to be phenomenalist because he wanted to found a science that was based purely on the results of observations and the mathematical correlations among these observations. In other words, he preferred laws to theories. This epistemological outlook ensured the early success of biometrics but, at the turn of the 20th century, Pearson's approach encountered resistance from the University of Cambridge where another element of the evolutionist paradigm was dawning: Mendelian genetics.

## Bateson and mutationism

At the end of the 19th century, the sudden, discontinuous variations that Darwin had called "sports" or "single variations", and that had been the bone of contention in his conflict with Jenkin, were confirmed by several authors including a Cambridge zoologist, William Bateson (1861-1926). Bateson was a follower of Galton and, like him, supported the idea of discontinuous variations. In 1894 he published *Materials for the Study of Variation Treated with Special Regard to Discontinuity in the Origin of Species*. Bateson thought that new living forms appeared through a discontinuous process, by a jump or series of jumps. In order to conserve changes, heredity must also operate in a discontinuous, non-blending manner. This was Mendel's idea thirty years earlier; it was "rediscovered" by three botanists in 1900.

Since 1886, the Dutch botanist Hugo de Vries (1848-1935) had been working on a cultivated plant from America, *Oenothera lamarckiana*, in which he had observed sudden, discontinuous variations. He drew analogies between his data and observations on other plants and animals,[34] and published an important two-volume (1901 and 1903) opus, *Die Mutationstheorie*. He called the sudden, hereditary variations which he considered to be at the origin of speciation, mutations. We now know that many of the variations that de Vries called mutations were in fact chromosome recombinations. Because of the special chromosome conformation of *Oenothera lamarckiana*, these recombinations produce stable forms. What we call mutations today were too few in number for him to observe in his work-samples. This should not deter us, however, from considering de Vries' work as a turning point. Forty years after Mendel had shown that the parent type was conserved from one generation to the next and had seen this as a confirmation of fixism, the science of genetics (name given by Bateson in 1906 to this new science of heredity) introduced the idea of mutation.

As mentioned above, if we think that mutations are important in the evolutionary process, then we embrace a view of heredity that precludes blending. In 1900, de Vries and two botanist colleagues, the German Carl E. Correns and the Austrian Erich von Tschermak, discovered, simultaneously but independently, the principles governing the hybridisation of plants that had already been established by Mendel. William Bateson followed by others, including Lucien Cuénot in France, extended these principles to animals. They became known as Mendel's laws and, ever since, we speak of Mendelian characters and call the science of cross-breeding Mendelism.

Once the notion of mutation had been integrated into Mendel's laws, transformism and genetics became compatible. However, this led Bateson and the first geneticists to reject Darwinism and natural selection. For them, the transformation of species was due to mutations and not to natural selection.

**Goldschmidt's "hopeful monster**

The idea that species can be transformed through mutation is present in the "hopeful monster" concept proposed by Richard B. Goldschmidt. In 1940, in *The Material Basis of Evolution*, Goldschmidt postulated that two kinds of genetic events might play a role in evolution: macromutations that have a substantial effect on the phenotype and systemic mutations that lead to an in-depth change in the genotype. In both cases, the new phenotypes are so different from the standard model of the species that they can be called monstrous. Most of these monsters have no future but Goldschmidt postulated that some survive and initiate an entirely new lineage. Nowadays, this phenomenon is thought to be extremely rare although not impossible. Goldschmidt's view has fired the imagination of non-scientific circles in Western culture. Our science-fiction is alive with "mutants", some said to guarantee the future (or salvation) of the species.

## The debate turns nasty

The two camps, Cambridge and London, showed no pity toward each other. Weldon and Pearson attacked Mendelism criticising its *ad hoc* character (to explain how traits vary, we need only assume an appropriate series of Mendelian factors); the Mendelians stressed how poor was the biology behind the mathematics. However,

> *as usual, the generals hated each other much more than the soldiers did, the soldiers being in the event the young researchers who were working under the guidance of Bateson and Pearson and who were not as temperamental intellectually as their elders. But, because of the clout of the leaders such as Pearson, Weldon and Bateson, the outside world could only come to the conclusion that Mendelism and biometrics were totally incompatible.*[35]

In his critical appraisal of biometrics, the Danish botanist Wilhelm Johanssen (1857-1927) was the first to use the term gene. He also introduced the terms genotype and phenotype. Genotype denotes the entire gamut of gene forms (or alleles) that occupy the chromosome locus (or loci) under study in an individual; phenotype denotes the expression of the genotype in a given environment. In one of his studies, Johanssen examined how the weights of seeds vary in a population of beans, *Phaseolus vulgaris* (see Fig. 2.2). Johanssen took 19 beans from a bag; their weights were distributed at the two ends of the range. He established a pure line from each bean and noted that selection had no effect on the mean weight of the seeds. The mean weight in each line was constant whether he selected the smallest or largest seed at each generation. It was clear that Johanssen's conclusion refuted the biometricians' thesis; natural selection did not work for continuous characters,

> *the action of selection cannot go beyond known limits - it must stop when the population, or, in practice, the isolation of the most strongly divergent pure line is complete.*[36]

The full explanation for this will only become clear much later. The bean is autogamous, and consequently selection does not operate within a family. Variations in characters are due to growth factors and environmental effects but not to genetic variations. The initial selection (the choice of 19 beans) only worked because Johanssen's beans were genetically different.

Pearson did not miss the chance to criticise Johanssen's work. In an article published in 1904 that was meant to refute Mendel's theory, he used the theory to explain continuous variations and the correlations that are observed between parents and offspring for this kind of variation. In his *reductio ad absurdum*, Pearson supposed that continuous variations were under the control of several genes, each character being completely dominant. This could explain both the normal distribution and the correlation between parents and offspring but the predicted correlation was 1/3 and not 1/2 as observed. Pearson's demonstration led to the widespread belief that Mendelism was refuted definitively but in 1909, he admitted that his reasoning had flaws. Different levels of dominance and different proportions of alleles in the population could account for the observed variations. The rift between Cambridge and London, Mendelism and Darwinism, mutationism and biometrics was still wide open.

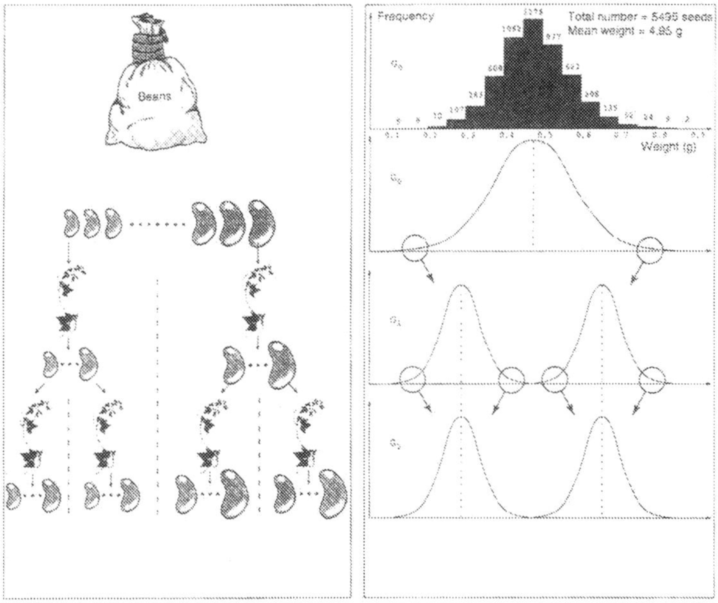


*Figure 2.2.* To refute the view held by biometricians, the Danish botanist Wilhelm Johanssen tested the effect of selection on a continuous character, the weight of beans (*Phaseolus vulgaris*). He took from a bag of assorted beans (left) 19 beans whose weights were at the two ends of the overall range. On the right is the normal distribution of the beans in the whole bag. Each of the 19 beans thus obtained, once self-fertilised, yielded seeds (G1) with a mean weight that was a function of the weight of the starting bean. (We show only the plants obtained from the smallest and largest of the 19 seeds.) Over the next generations (G2 and seq.), however, selection no longer affected the distribution of the seed weights. The variation in these pure lines, due to accidents in growth, is thus not inherited.

The work of the American geneticist Thomas Hunt Morgan (1866-1945), which followed on from de Vries' research on heredity and chromosome theory, did not bridge the gap. Morgan was nominated professor of biology at the University of Columbia in 1904. He chose to study the fruit fly, *Drosophila melanogaster*, that can be bred in thousands, millions of exemplars. At first, Morgan tried to demonstrate that mutations were sudden variations due to the environment. He subjected flies to traumatic physical conditions and to chemical agents but induced virtually no mutations, and these had no adaptive value. The emergence of a mutant male with white eyes prompted him to make countless crosses with normal flies. His conclusions were similar to Mendel's and he converted to Mendelism. Using the "giant chromosomes" of the fruit fly's salivary glands which are easy to observe with a microscope and which have characteristic bands, he established chromosome maps. Morgan did not only prove the physical reality of hereditary factors but also localised them! The chromosome theory of inheritance was thus established definitively. Together with his collaborators, Calvin Bridges, Alfred Henry Sturtevant and Hermann Joseph Müller, he published *The Mechanism of Mendelian Heredity* in 1915, *The Physical Basis of Heredity* in 1919, and *The Theory of the Gene* in 1926. He also studied chromosome crossing-over whereby parts of homologous chromosomes are exchanged.

Morgan and the geneticists continued to fight the Darwinians. They argued that the function of natural selection was not to be creative but to eliminate or maintain discontinuous variations which were the real source of evolution. Consequently, useless or harmless new characters could appear and be maintained in populations, in the absence of selective pressures. Differences in species could thus concern characters of minor importance that apparently do not derive from an adaptative process.

As explained in the next chapter, the two approaches will be reconciled in the neo-Darwinian melting-pot, chiefly thanks to population genetics. However, before broaching the neo-Darwinian paradigm - which is the basis of current models of evolution - we have to mention two socio-political avatars that this new outlook on life brought with it, or had inflicted upon it. These avatars were influenced by earlier views mentioned in this chapter.

## 4. EUGENICS AND LYSENKOISM

### From eugenics to "eugenism"

Eugenics was introduced as a discipline around 1910 chiefly in Anglo-Saxon countries (including Germany) and France. Francis Galton had already used the word in 1883 to denote the study of socially controllable elements that can improve or impair, physically or mentally, the racial qualities of future generations. What was the reason for Galton's interest? The historian Bernard Norton supports the hypothesis put forward by Ray Fancher and Daniel Kevles that goes as follows. Although Galton's family took him for a genius, he failed his Tripos exams at Cambridge University. This failure induced a passionate curiosity for the origins of intellectual success and a no less passionate desire to explain his own lack of success. After reading *The Origin of Species*, he came to the conclusion that:

> *a man's natural abilities are derived by inheritance, under exactly the same limitations as are the form and physical features of the whole organic world. This explained the limits of his own abilities*[37]

and yet opened up a whole new world before him. Whether this explanation is true or not - we shall leave the responsibility for it with Norton - Galton intended to imitate nature using the human species as his material. His plan was to bring society forth out of the age of Christianity into the age of post-Darwinism, that is into the age of being "well-born", of orchestrated improvement, in other words, of eugenics. To this end, he founded in 1907 the Eugenics Education Society which had as one of its directors Charles Darwin's son, Leonard Darwin. The Society published the *Eugenics Review*.

**Eugenics is based on two fundamental postulates:**

1. Genes are at the origin of differences in the potential and behaviour of individuals; they are subject to evolutionary pressures.
2. The human species is the king pin of evolution; it owes its status to natural selection which needs to be helped along in order to maintain this status.

Mankind did not wait until the beginning of the 20th century to battle against the laws of nature. The art of healing, customs, social laws (in particular, aid to impoverished families), and culture are all ways of influencing the course of natural evolution. But geneticists ask, is this battle always for the "good" of the human species? Should we not show more concern for the genetic heritage of the human species and protect it against degradation? Is it not time to offer mankind the means to control the evolution of its own heritage? This was the aim of the science of eugenics, to assist and even induce the birth of "good", i.e. better, individuals and populations.

In a lost novel, Galton described an island governed by a eugenic council whose goal was to enhance the biological quality of the inhabitants. To be admitted to the island, immigrants had to pass a test. The results of the test determined the number of children they were allowed to have. Authority was in the hands of members of the university. The guiding principles of Galton's utopia were individual biological quality and scientific power.

There is no reason to doubt the good faith and philanthropic attitude of the geneticists who developed positive eugenics, as it will be called later. This form of eugenics is confined to the study of how to improve the reproduction rate of the most gifted. It calls upon many scientific disciplines which are seen by the adepts of eugenics as the strong roots feeding a harmonious entity. It is a totally different matter, however, when scientists who claim to be building Galton's utopia usurp all human scholarship and, in the name of science, take control over the civil and political dimensions of society. In France, we then no longer speak of eugenics but eugenism. Its limits are not as readily defined. Eugenism includes the promotion of negative eugenics, i.e. limiting the number of least fit offspring, and is not unrelated to social Darwinism.

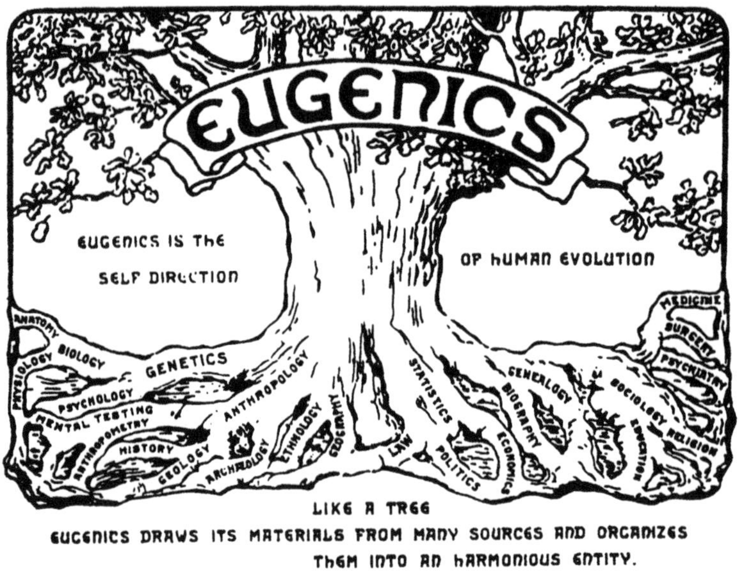


*Figure 2.3.* Eugenics

The expression social Darwinism was introduced in 1880 to cover theories in which social laws are more or less direct extensions of the laws of nature (in particular the "survival of the fittest" or the "struggle for life").

> *To the mind of social Darwinians,* notes Daniel Becquemont, *interracial conflict drives history, and eugenist policy is part of this conflict. Even if many eugenists have severe doubts about absence of immigration control and may express latent or outright racist views, they nevertheless believe that new gifted elements from other races are beneficial; in particular, they often consider that controlling the racial quality of a population is a way of substituting the more civilised practice of artificial selection for the inevitability of interracial conflict.*[38]

Whether eugenism is linked to social Darwinism or not depends on whether it chooses to support the idea of conflict between races or selection among races.

The press, and especially the North American press, often relied on scientific statements and data to back eugenist views. Indeed the public would do well to know about some of the data! Who could have imagined that Ada Jukes, born in 1740, would be at the root of an absolutely catastrophic family tree that included 64 mentally diseased offspring, 174 sex-perverts, 196 illegitimate children (were they really her descendants?), 142 paupers (an absolutely unacceptable condition at the time), 77 criminals and murderers. The newspaper's only comment was:

> *Shall we allow the Ada Jukes of today to continue this multiplication of misery?*

... and we might add of crime and perversity!

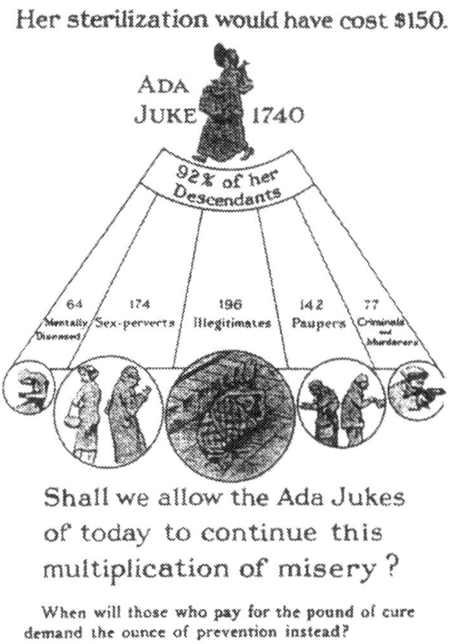


*Figure 2.4.* Ada Juke's family tree.

Although the Sunday Oregonian is neither a scientific nor a medical journal, it nevertheless took an interest in the health of the offspring of a couple of idiots, Karl and Anna. Karl and Anna's family tree was most demonstrative. It included not only a disturbing number of feeble-minded individuals, but many of these were not even interned in psychiatric wards! No doubt, the conclusion the newspaper wanted to impress upon its readers was that those who were performing sterilisations in Oregon should keep going. There is even a case of disjunction of an illegitimacy gene that was reported by an unpretentious local American newspaper. This gene was present in four generations: Maria, the grandmother, Ruth, the mother, Lena, the daughter, and William who was just 2 weeks old and already showing sure signs of being a carrier of the illegitimacy gene (that is, if we believe the journalist).

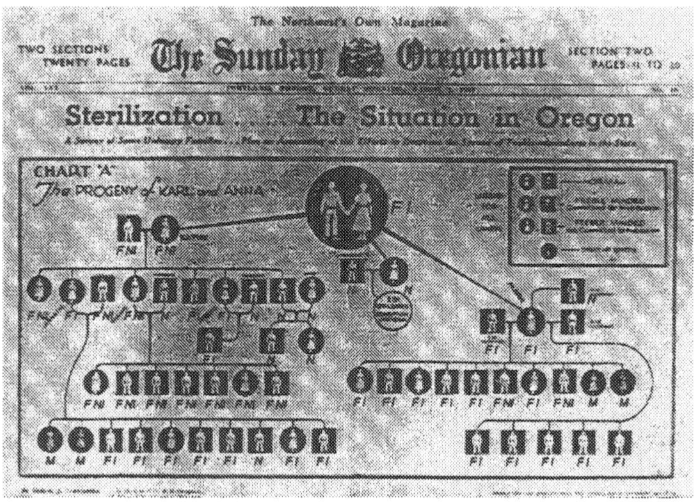


*Figure 2.5.* "Sterilisation. The situation in Oregon. A survey of some unhappy families ... plus an accounting of the efforts to decrease the spread of feeble-mindedness in the State." In the article, the categories of Karl and Anna's offspring were distinguished by colour (not visible in black and white). We have added letters to indicate the numbers of individuals who were normal (N), stillborn (M), feeble-minded and interned (FI), feeble-minded and not interned (FNI).

As if the newspapers weren't enough, the American Eugenics Society set up stands to explain that the United States needed more "good" people (of whom few are born) and less "mediocre" or "bad" people (who are unfortunately born in large numbers). The results of this campaign were quickly felt. Laws inspired by these eugenic currents were passed and sterilisation programmes were set up. By January 1, 1935, 21 539 people had been sterilised in America because they had "bad genes". The laws were not abrogated until fairly recently (in 1972 in the State of Virginia).

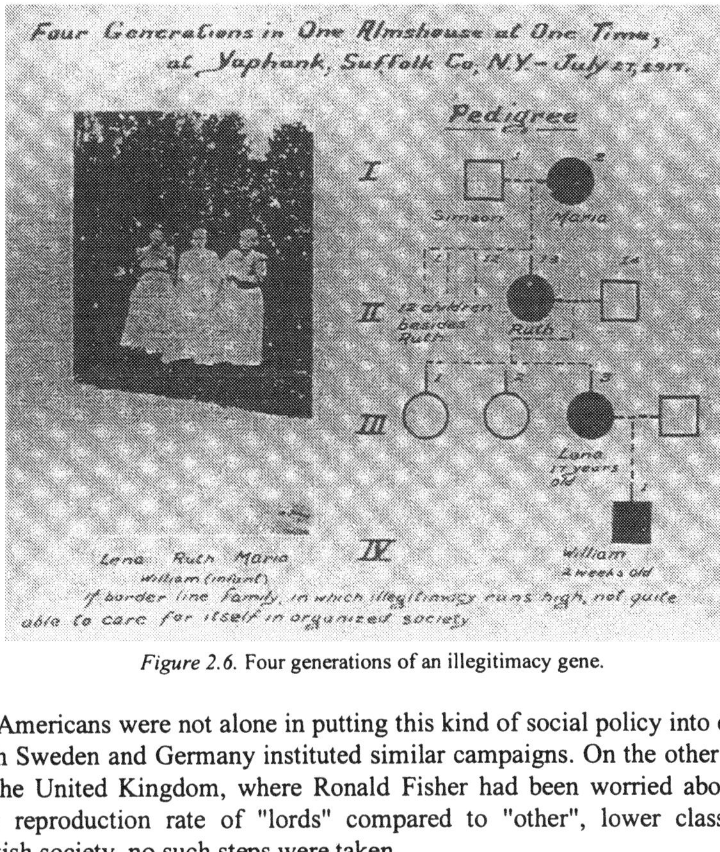


*Figure 2.6.* Four generations of an illegitimacy gene.

Americans were not alone in putting this kind of social policy into effect; both Sweden and Germany instituted similar campaigns. On the other hand, in the United Kingdom, where Ronald Fisher had been worried about the low reproduction rate of "lords" compared to "other", lower classes of British society, no such steps were taken.

Scientists dissociated themselves fairly quickly from this enterprise especially after the monstrous turn it took in Nazi Germany. The Germans had imitated the hard-line American eugenists and even vexed them by claiming larger sterilisation numbers. However, there was a crucial difference between the Americans and Germans. The Americans sought to improve the race by targeting individuals carrying bad genes; the Germans based their undertaking on whether a person belonged to a particular race. This distinction should be kept in mind when we analyse current statements and projects. Moreover, we should not forget how readily scientists of the first half of the 20th century took part in a movement that seems so horrible and reprehensible to us now. They were no doubt as aware of their responsibilities as today's scientists and also convinced that they were acting for the good of mankind. Let's take heed of the traps and snares of science.

What has become of eugenism today? Did it really come to an end with World War II? Opinions are divided. Nazism has induced such an intense state of shock that attempts to improve the human species have been forsaken. The genetics on which first eugenics, then eugenism was based was of such poor quality that it has been abandoned by research workers and adepts. However, the risks of a return to eugenism have probably never been greater than they are today, in two respects. First, there is the issue of birth control which is in some ways related to social Darwinism. Admittedly, the planet's population explosion is worrying but how should we view the ways in which birth control is thrust upon Third World countries? The motives for birth control programmes do not always lie in pure philanthropy but also arise from a need to maintain or impose a certain social and political balance. Second, there are the recent developments in genetics and molecular biology. What should we think of the techniques and work programmes for decoding the human genome, screening for hereditary diseases, selecting embryos or investigating gene therapy? Of course, we must distinguish between the personal reasons for which parents might use these methods and the totalitarian rules of a State that decides which individuals may reproduce and survive within a population. However, the enthusiasm generated by these techniques, and arising from highly effective orchestrated media publicity, does not create a climate that awakens individual consciousness and responsibility.

French public opinion rejects the conclusions of sociobiology, sometimes considered as the heir of social Darwinism, but paradoxically rarely questions the eugenic temptations of modern medicine or asks at which level (individual, family, society, nation, etc) should responsibility lie.

During the last fifty years, biology has not come up with a definition of man that facilitates medicine's art of healing and devotion to the "good" of mankind. How should we view alterations (for better or for worse) of a human being's genetic inheritance today? Can one always differentiate between the search for mankind's "well-being-ness" and "well-born-ness"? The eugenics-eugenism file has taught us at least two lessons to be put into immediate practice:

1. It is part of a research worker's code of ethics to ensure that statements made by scientists, or by others on the basis of a scientist's work, are exact. This is essential as society has great faith in the scientific word.

2. The theory of evolution in Darwin's day, but even more so in our modern genetic engineering age, seems to provide a vision and an understanding of the future. It is tempting to want to do things for the good of future as well as present generations, in the name of scientific knowledge

and with the help of this knowledge. Is this one of the social roles of the scientist? A moot point indeed.

## Neo-Lamarckism and Lysenkoism

The first thirty years of the 20th century witnessed the birth of many theories attempting to explain evolution: August Pauly's psycho-Lamarckism (1905), Hans Driesch's entelechy, Daniele Rosa's ologenesis (1909), Smuts' holism (1916), Henry Fairfield Osborn's aristogenesis, Lev Semenovich Berg's nomogenesis (1922), Labbé's allelogenesis (1924), Ludwig von Bertalanffy's general systems approach (1928) and Przibram's apogenesis (1929)! [39] Neo-Lamarckism had a great following in France with well-known adherents such as Giard, Le Dantec, Perrier, Bonnier and Houssay although inheritance of acquired characters, which many saw as the central dogma of Lamarck's theory, was still unproven. In 1909, the Viennese scientist Paul Kammerer claimed to have established inheritance of acquired characters in the midwife toad but he was accused by Bateson in 1923 of correcting nature. The partisans of neo-Lamarckism contended that the genesis of certain structural characters that were ill accounted for by the Darwinian paradigm could be easily explained by the effects of use and disuse. Examples were the callosity of the camel, ostrich or wart-hog, the camber of the hermit-crab's abdomen, and so on. In the middle of the century, Lamarckism resurfaced with Paul Wintrebert's (1867-1966) hypothesis of a mechanism not unlike immunisation. The gene was an antibody bound to the chromosome and thus a kind of reaction against environmental disturbances. It was not heredity's agent but instrument. The addition of genes (adaptive responses) determined the evolution of species.

Toward 1928, the Russian Trofim Lysenko (1898-1976) built his reputation on a procedure called vernalisation. It was based on the observation that winter cereals can give rise to heads even if sown in spring, as long as the seeds, once they have been humidified and have germinated, are kept at relatively low temperatures. He used this procedure, which belongs to plant physiology and not genetics, as proof that we can modify heredity without influencing population structure. Lysenko's inability to distinguish between "training" an individual - even when it is reduced to a seed - and genetically modifying a population can be linked to Lamarckism. It led him to believe that nature can be trained. However, what at first seemed to be just another episode in the history of northern agronomy soon became the "Lysenko affair",

> *the perfect model of the construction of a false science by combining a dogmatic ideology (Marxism-Leninism) with the totalitarian inquisition of a dictatorial system (the Soviet communist state). It was the triumph of*

*a pseudo-science over a genuine science - classical genetics - that was forbidden by ukase to the soviet peoples and the colonies of the Kremlin's empire.*[40]

Lysenko did not even discover vernalisation; he just had the idea of applying it to agronomy. Whether he succeeded in this venture (as Dominique Lecourt and Francois Dagognet believe) or not (Denis Buican's view [41]), Lysenko managed to impose his ideas on the regime and society of the Soviet Union. Lysenko claimed that he was following in the footsteps of the arboriculturist Ivan Mitchourin (1855-1935). Mitchourin had produced hybrid grafts which did not seem to obey the rules of classical genetics based on Mendel's and Morgan's discoveries. He was the son of a small outcast landowner who was an employee of the railways and then a watchmaker. Fascinated by horticulture, he bought an orchard in 1875 which was later to become an experimental station, the Flower and Fruit Selection Centre, and then the "Ivan Vladimirovitch Mitchourin" Central Genetic Laboratory. Mitchourin had little time for theory (the doctrines of Mitchourism were enunciated by Lysenko); his approach was empirical. He believed that mankind could modify nature and control its development, including the processes of heredity. Heredity was defined as "*the property of the living body to react in a certain fashion to environmental conditions*" and then to pass on this change to the next generation by way of any somatic particle (and not only the chromosomes). Heredity was similar to metabolism ... and could therefore be modified by it. The proof was vegetative hybridisation, the name Mitchourin gave to the theory in which support and graft mutually influence each other and where inheritance occurs via the sap. Mitchourin was not hostile to Mendelism as such and used the term gene now and again, but he doubted whether the laws of inheritance could be established *a priori* and whether the results of interbreeding were as predictable as the predictions made by Mendel with peas.

Lysenko was even more categorical. He claimed to have adapted plants to specific conditions by vernalisation. From 1930 onwards, at the Institute of Plant Genetics and Reproduction in Odessa, he undertook physiology experiments, especially the production of graft hybrids, in order to show that "plastic matter" circulating between the support and graft could transform support or graft into a more or less intermediate type by inheritance. The graft's influence on the germ cells of the support was reflected in the transmission of the characters of form and colour to the support's offspring. These were thus characters that were acquired and inherited.

Lysenko's rise in society (he called himself an agrobiologist) began in 1935 [42] when he gave a speech at the Collective Farmers Congress which embraced vernalisation and the class war in a single breath. His speech was rewarded by a "Bravo" from Stalin which was echoed in Pravda. At this point, Soviet biology officially entered the era of Lysenkoism. Together with the philosopher and Marxist-Leninist academy member Prezent, Lysenko attacked chromosome theory which he said was "*reactionary, idealist, metaphysical and sterile*". In 1936, when he presented a paper entitled *Challenging the Problems of Genetics and Breeding* at the Lenin Academy of Agronomic Sciences, the rift between classical genetics and Lysenkoism was wide open. The VIIth International Congress of Genetics which was to be held in Moscow the following year had to be transferred to Edinburgh in 1939. In the meantime, Lysenko climbed the ranks of office and honours. He became president of the Lenin Academy of Agricultural Sciences, director of the Genetics Institute of the Academy of Sciences (which was to be called after him in 1948), was a beneficiary of the Stalin prize, a hero of the USSR, *etc.* His main rival, the academy member Vavilov, one of the founders of plant genetics, was arrested in 1940 and sentenced to death. The sentence was changed to life imprisonment but he was murdered in his cell in 1943. All geneticists unable to leave the Soviet Union met with a similar fate.

After the war, Lysenko continued to promulgate his ideas, pointing time and again to the graft hybrids. Opposition was qualified as "*reactionary Mendelism-Morganism*" and tracked down relentlessly. Soviet biology had become Lysenko's monopoly; its heyday was between 1948 and 1952. The results of his fraudulent experiments on the transformation of one species into another (wheat into rye, oats into barley, cabbage into swede, pine into fir) were widely publicised. This is not the place to dwell on the misery of Soviet agriculture at the time but as D. Buican says:

> *Lysenko with his harmful "scientific" advice only added a few drops to the ocean of dearth that bathed the Gulag archipelago.*[43]

Contrary to expectations, Stalin's death did not sound Lysenko's decline. He quickly found other political support and, thanks to Krustchev, triumphed again between 1962 and 1965.

> *That's what science is about [speaking of Lysenkoism]. As you can see, Lysenko had already introduced a positive point into the debate which many comrades pursued at this session. Anyone willing to adopt Lysenko's methods cannot lose. Go and have a look at his wheat; go this year. I am sure that he will always have a good harvest to show you. Go and have a look at the corn on his farm, at his sugar-beets... This is the type of scientist who has something to teach us.*[44]

He was not dismissed from his post as director of the Genetics Institute until 1965. The following year marked the return of classical genetics in the Soviet Union.

**NOTES**

1. Denton, M., 1985, *Evolution: A Theory in Crisis*, Burnett Books, London, p. 358. Michael Denton is not a creationist as is sometimes thought.
2. Denton, M., *ibid.*, pp. 358-359.
3. Popper, K., 1972, *Objective Knowledge: An Evolutionary Approach*, Clarendon Press, Oxford University Press, London, p. 267.
4. Lecourt, D., *L'Amérique entre la Bible et Darwin.* Series: Science, Histoire et Société, Presses Universitaires de France, Paris, p. 114.
5. Kuhn, T. S., 1970 (first published in 1962), *The Structure of Scientific Revolutions* (2nd edition, enlarged), The University of Chicago Press, Chicago, p. 10.
6. Quoted by Taylor, G.R., 1963, *The Science of Life. A Picture History of Biology.* Thames & Hudson, London, p. 151.
7. Lamarck, J.B., 1994 (originally published in 1809), *Philosophie Zoologique*, Flammarion, Paris.
8. Darwin, C., 1859, *On the Origin of Species by Means of Natural Selection*, Murray, London.
9. Gayon, J., 1992, *Darwin et l'Après-Darwin. Une Histoire de l'Hypothèse de Sélection Naturelle,* Kimé, Paris, p. 1. (transl. *Darwinism's Struggle for Survival: Heredity and the Hypothesis of Natural Selection* (Cambridge University Press, Cambridge, 1998.)
10. Gilson, E., 1971, *D'Aristote à Darwin et Retour. Essai sur Quelques Constantes de la Biophilosophie,* Series: Essais d'Art et de Philosophie, Vrin, Paris, p. 87.
11. Darwin, C., 1859, *ibid.*, Chap. II.
12. Darwin, C., *ibid.*, Chap. II.
13. Darwin, C., *ibid.*, Chap. III.
14. Darwin, C., *ibid.*, Chap. IV.
15. Quoted by Taylor, G.R., *ibid.*, p. 157.
16. Quoted by Taylor, G.R., *ibid.*, p. 157.
17. Gayon, J., *ibid.*, pp. 60 & 61.
18. Quoted by Taylor, G.R., *ibid.*, p. 165.
19. Gayon, J., *ibid.*, p. 106.
20. Darwin, C., *ibid.*
21. Darwin, C., *ibid.*
22. Mayr, E., 1985, Weismann and Evolution, *J. Hist. Biol.* 18: 311. Mayr points out that Weismann added many notes on a variety of subjects to the English edition of *Studien zur Descendenztheorie* (1876) which was published in 1882. However, he did not restrain the blatantly Lamarckian statements of the original German text.
23. Weismann, A., 1887, *Über die Zahl der Richtungskörper und über ihre Bedeutung für die Vererbung,* Fischer, Iena.
24. He was a student of Christian Doppler who described and explained the well-known "Doppler effect", i.e., the frequency shift that occurs on reception of waves emitted by a source that moves with respect to the observer.
25. Blanc, M., 1984, Gregor Mendel: la légende du génie méconnu, *La Recherche* 15: 48.
26. Quoted by Taylor, G.R., *ibid.*, p. 312.
27. Gayon, J., *ibid.*, p. 287.
28. Gayon, J., *ibid.*, p. 290.
29. Quoted by Blanc, M., *ibid.*, pp. 58-59. The letter is reproduced in Orel, V. and Armogathe, J.R., 1985, *Mendel, un Inconnu Célèbre, Belin*, Paris, pp. 129-139.
30. Norton, B, 1981, La situation intellectuelle au moment des débuts de Fisher en génétique des populations, *Revue de Synthèse* 3 (N°s 103, 104): 241.
31. Galton, F., 1869, *Hereditary Genius: An Enquiry into its Laws and Consequences*, Macmillan & Co, London, p. 26.

32. Galton, F., 1885, Regression towards mediocrity in hereditary stature. *J. Anthropol. Inst. of GB and Ireland* 15: 246 & 247.
33. Pearson, K., 1900, On the criterion that a given system of deviations from the probable in the case of a correlated system of variables is such that it can be reasonably supposed to have arisen from random sampling. *Philosophical Magazine*, 50 ($5^{th}$ series): 157-175.
34. For example, observations by the Russian Sergei Ivanovich Korginski (1900, *Hétérogénèse et évolution. Théorie de la formation des espèces*) in whose mind the origin of the transformation of species resided in this type of variation.
35. Norton, B., *ibid.*, p. 238.
36. Johanssen, W, 1903, *Heredity in Populations and Pure Lines. A Contribution to the Solution of the Outstanding Questions in Selection.* ("Ueber Erblichkeit in Populationen und in reinen Linien"), Gustav Fisher, Iena. (Reprinted in 1959 in *Classic Papers in Genetics* (ed. J.A. Peters), Prentice Hall, Englewood Cliffs, p. 21).
37. Galton, F., 1869, *Hereditary Genius: An Enquiry into its Laws and Consequences*, Watts & Co, London, ($2^{nd}$ ed) p. 1.
38. Becquemont, D., 1996, Darwinisme social, in: *Dictionnaire du Darwinisme et de l'Evolution*, Vol. 1, Presses Universitaires de France, Paris, pp. 1415-1416.
39. Taton, R., (ed.), 1964, *Histoire Générale des Sciences. Part III: La Science Contemporaine; Vol. II: Le XXe Siècle*, Presses Universitaires de France, Paris, p. 714.
40. Buican, D., 1988, *Lyssenko et le Lyssenkisme*, Series: Que sais-je? N° 2394, Presses Universitaires de France Paris, p. 3.
41. Buican, D., 1984, *Histoire de la Génétique et de l'Evolutionnisme en France*; Chapter 8: L'Anti-génétique Soviétique en France, Presses Universitaires de France, Paris; Buican, D., 1987, *Génétique et Pensée Evolutionniste*, Chapter VII: Pseudo-sciences dans l'Histoire des Sciences: l'Affaire Lyssenko) and Chapter VIII: La Résonance Perpétuelle du Lyssenkisme), SEDES, Paris.
42. Medvedev, Z. A., 1969, *The Rise and Fall of T. D. Lysenko*, transl. I. M. Lerner, Columbia University Press, New York.
43. Buican, D., *Lyssenko et le Lyssenkisme*, p. 19.
44. Khrutchev, N., 1964, quoted by Buican, D., *ibid.*, p. 102.

# Chapter 3

# POPULATION AND ECOLOGICAL GENETICS

Neo-Darwinism arose from a need to reconcile Darwinism and Mendelism. The data on the transmission of hereditary material, obtained mostly from experiments on the fruit-fly from the beginning of the 20th century onwards, was increasingly precise and had to be made to fit in with the theory of evolution. The mathematician Ronald A. Fisher and the geneticist Sewall Wright were the first in the attempt. In the wake of the biologist J.B.S. Haldane, they built models that described the evolution of populations and founded population genetics. The early models were based on pure speculation because the values of most variables were unknown. To fill this gap in knowledge, Edmund B. Ford created ecological genetics (evolutionary ecology); its aim is the experimental study of populations in their environment.

## 1. POPULATION GENETICS

How do species evolve? At the beginning of the 20th century, Darwinians claimed that species evolve through the action of selection whereas Mendelians insisted on the role of mutation. We now know that both selection and mutation are involved but it took us thirty years (i.e. until the 1930s) to come to this conclusion. Mutation introduces new alleles[1] (new gene forms) but, once the alleles are there, what happens to them? Well, it is at this point that selection intervenes and, to understand in what way, a new concept, allele frequency, was introduced. If a gene can mutate into a new form, we must be able to evaluate what proportion of the population will carry this new allele and how this proportion changes with time. This is what population geneticists do.

## Evolutionary pressures

Population geneticists not only build mathematical models but also observe natural populations and perform laboratory tests to measure the variables in their models. Let us suppose that the individuals of a population have either gene form *A* or *a* at a given locus of their genome, and let $p$ and $q$ be the proportions of these two alleles within the population ($p + q = 1$). The models of population genetics describe how $p$ and $q$ change with time. (Some scientists believe that evolution is nothing other than the sum total of the frequency changes of all genes). The forces responsible for the change in the allele structure of populations are known as evolutionary pressures. We want to model their action.

> **"Wild-type allele"**
>
> The reader should beware of the expression "wild-type allele" because, in natural populations, there are often several different alleles for a given gene (locus). This undermines an idea from which we should steer clear, namely, that there are "normal" types from which certain individuals (mutants, *par excellence*) deviate.

Let us take the simplest case of so-called haploid species whose cells have just one copy of each chromosome. (For example, bacteria are haploid and have a single, circular chromosome). In a population of haploid organisms, gene forms *A* and *a* occur nA and na times respectively, where n is the number of individuals carrying the allele. Usually, instead of finding out how nA and na change over time, we study the changes in the overall population (N = nA + na) and in the frequency (p = nA/N and q = na/N = 1-p). Studying variations in N is part of demography and in p of population genetics. In population genetics, N may be a variable that influences the variations in p but is not a variable whose variation is explained. The different approaches of demography and population genetics set the two disciplines somewhat apart. Population genetics would be hard put to explain the effect of genetic composition on the demographic growth of a species or subspecies. This is one of its limitations; we shall mention others later. Its role is to describe the changes in genetic composition under the action of four evolutionary pressures: mutation, selection, migration, and genetic drift (see annex MA1 for a mathematical analysis of these pressures).

## Mutation

If by mutation we mean any hereditary variation in genetic information, then many events give rise to mutations. Most mutations are errors in the reproduction of the genetic message:[2] simple copying errors (one base replaced by another), stuttering of the replication machinery (sequence duplication), and omission (deletion). Errors may also occur during more complex phenomena such as DNA repair and DNA recombination (transposition[3], gene conversion), chromosome rearrangement (translocation, inversion), or cell division (mitosis or meiosis) when an entire chromosome (aneuploidy) or a whole set of chromosomes (polyploidy) is added or deleted. Population geneticists study all these mutations. For each type, they need to know the probability of its occurrence at each generation and per unit of time in order to calculate its impact on allele frequency.

### DNA

DNA (deoxyribonucleic acid) is the well-known molecule with a double helix structure in which heredity is inscribed or, in Jacques Monod's words, the "*fundamental biological invariant*".[4] The Swiss chemist Friedrich Miescher isolated DNA from salmon sperm in 1869 but it was not until 80 years later that the molecule was identified as the hereditary substance of chromosomes. Several advances occurred in the intervening years: the biochemical mechanisms that control gene expression began to be investigated experimentally; G. Beadle and E.L. Tatum's experiments on amino acid biosynthesis established an unequivocal relationship between gene and enzyme; this was followed by the work of Fred Griffith and then of Oswald Avery on the virulence of pneumococci and their transforming principle (shown to be DNA); their results were soon to be criticised by Max Delbrück and the "phage group"; Linus Pauling discovered the alpha-helix motif in the structure of proteins but failed to understand the structure of DNA, and Edwin Schrödinger showed astonishing prescience in his essay "*What is life?*". These and other milestones finally led to the publication in 1953 of James Watson's and Francis Crick's paper on the so-called "double helix" model. Francois Gros sees this model as the emblem of molecular biology, just like Mercury's wand (Aesculapius' staff) is the symbol of the art of healing.[5] In 1981 Watson wrote:

*But ... we did not foresee any immediate practical consequences for the world about us and no reason why the man on the street, as opposed to the young student, should know that we existed, much less try to understand the increasingly complicated facts that were to be worked out over the next twenty years.*[6]

The spontaneous mutation rate was measured by observing the spontaneous inactivation of mutated genes. It is fairly low in all species, about $10^{-6}$ mutations per gene per generation. At this rate, 700 000 generations are needed for mutation to halve the frequency of an allele. This is equivalent to about 10 to 20 million years for the human species and its forebears and takes us back to the tertiary era, well before we emerged! A force that needs so much time to just halve allele frequency can hardly have played a major role in the evolution of populations; its effect must be negligible compared to that of other evolutionary pressures. However, mutation does have a key evolutionary role because it produces new alleles. There is not a single gene in our entire genome that has not been produced by a series of mutations. It is only when mutation has introduced an allele into our hereditary pool that other evolutionary pressures intervene to influence allele frequency.

**Selection**

When Charles Darwin made the expression "natural selection" a mainstay of his theory of evolution, it had for him the common intuitive meaning that comes from a knowledge of artificial selection as practised by breeders:

*Can it, then, be thought improbable, seeing that variations useful to man have undoubtedly occurred, that other variations useful in some way to each being in the great and complex battle of life, should occur in the course of many successive generations? If such do occur, can we doubt (remembering that many more individuals are born than can possibly survive) that individuals having any advantage, however slight, over others, would have the best chance of surviving and of procreating their kind? On the other hand, we may feel sure that any variation in the least degree injurious would be rigidly destroyed. This preservation of favourable individual differences and variations, and the destruction of those which are injurious, I have called Natural Selection, or the Survival of the Fittest.*[7]

Later, Darwin's definition had to be adapted to take Mendelian inheritance into account and in order that it might be more quantitative. We now use the distribution curve of phenotypes in a population to distinguish three modes of natural selection:

1. Directional selection acts at one end of the curve (for example, a virus selects those individuals in a population which put up the best resistance);

2. Stabilising or conservative selection favours intermediate, median phenotypes (for example, selection acting on babies' weights at birth eliminates the extremes);

3. Divergent selection acts simultaneously at both ends of the distribution curve (for example, when a new predator arrives on the scene, the prey-species can increase in size for better resistance and decrease in size for better concealment).

Irrespective of its mode of action, selection acts on the phenotype which is determined by many genes and environmental factors. Because we cannot observe the evolution of the entire genome, we use the concept of the fitness of a genotype (also called adaptive or selective value) to study the change in allele frequency at each locus under the effect of selection. This change depends on the survival and reproduction capacities that the allele confers to the carrier. Thus, the fitness of a genotype is defined as the number of offspring the carriers of the genotype leave at the next generation. It is directly dependent upon viability (the probability of reaching adulthood) and fertility (the number of offspring an adult leaves behind).

Fitness depends on the interaction of the genotype under study with all the other loci of the individual's genome. We cannot therefore attribute observed differences in fitness just to different genotypes at the locus under consideration; they may be due to differences in other genes at other loci. Fitness also depends on the environment which can privilege certain genotypes. Finally, estimating viability and fertility is part of demography. To avoid random effects, we must compare many individuals with different genotypes. Some individuals in a population reproduce more than others for no genetic reason. All these studies belong to ecological genetics (see Section 2 of this chapter).

## Migration

Migration is the transfer of individuals or gametes between populations of the same species. It passes on and exchanges genes. These gene fluxes alter allele frequencies in populations. Migration is thus an evolutionary pressure. Described as *"a sort of genetic glue that holds populations together"*,[8] it increases the homogeneity of allele frequency in a set of populations. Its effects are noted:

1. on a population scale: migration limits differentiation of isolated populations originating from the same initial population. (This genetic differentiation is due to genetic drift - we shall come back to this below -, mutation and selection if the populations are subject to different selective pressures.)

2. on a local scale: migration runs counter to processes that decrease variability.

How can we define a population when there is migration? How can we divide species into populations? Don't individuals mate with close rather than distant partners? The boundaries chosen by geneticists are somewhat arbitrary. They define a population as the group that minimises the migration rate whilst preserving a certain homogeneity. In practice, unless the boundaries are clear-cut, the "population" under study will tend to be the ensemble that can be sampled in a reasonable time. There is the risk, however, that this unit will be smaller or larger than the one undergoing gene pooling at each generation.

## Genetic drift

Genetic drift is a well-established process in population genetics but it is not easy to grasp. It can call into question the simple conclusions we have just noted. When calculating the action of the selective forces on a gene in order to describe the change in its frequency, we assume implicitly that the law of large numbers applies. In other words, if the population is large enough, the observed value for allele frequency will hardly differ from the calculated (mathematically expected) value. However, for this to be true, that is, for no drift, the population must be infinite. An infinite population does not exist in practice and the law of large numbers therefore does not apply. The genes carried by the gametes giving birth to a new generation are just a sample of the genes of the previous generation. The frequency of an allele thus fluctuates randomly from one generation to the next, even in the absence of mutation and selection. The interesting point about this process is its Markovian nature; it has no "memory". Once the new generation is born, it forgets the value of the allele frequency in the previous generation. Thus, in the absence of other evolutionary pressures, allele frequencies fluctuate randomly. This is known as genetic drift. Its rate is inversely proportional to population size.

What is the long-term consequence of genetic drift? If no force keeps allele frequency constant, it fluctuates and can take on any value between 0 and 1. But, it cannot do so indefinitely. Sooner or later, it will reach 0 or 1 and be fixed once and for all. If it becomes 0, the allele in question will have totally disappeared in favour of another (or several other) allele form(s) at the locus under consideration. If it reaches 1, it will have eliminated all other allele forms. Genetic drift has been compared to the gait of a drunkard on a ridge. If he cannot walk straight, he will end up by falling from the ridge, either on one side or the other.

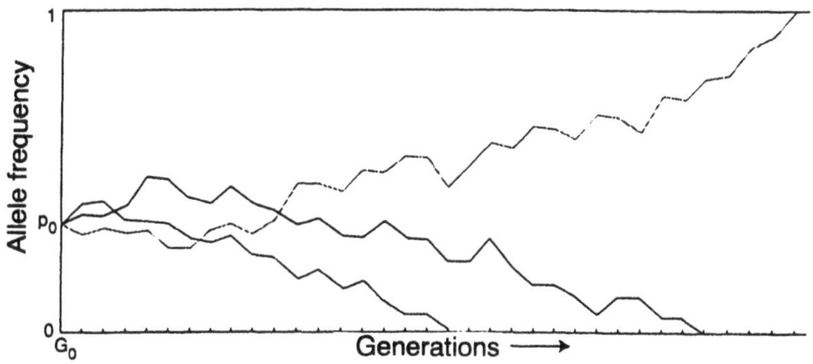


*Figure 3.1.* With time, genetic drift in a population leads to random variations in allele frequencies for each locus. The drift will only cease when the allele is the only one left or when it has disappeared. In the above figure the drift of an allele in a small population is simulated starting from an initial frequency p0 of 1/3.

Of course, selection can oppose genetic drift which remains present notwithstanding. If the population is large enough, the effects of drift will be negligible compared to those of selection. On the other hand, in a smaller population, selection is more hard pressed to oppose the effects of random drift. In other words, the weaker the selection force, the greater the populations have to be for selection to be effective. An allele that has only a small deleterious effect, for instance slight myopia, can thus invade a small population just by chance.

In the course of their history, some populations undergo an important reduction in size; one speaks of a population bottleneck. When this happens, a large part of allele diversity can be lost through drift and even become 0 for all loci.

The founder effect obeys the same principle but the scenario is reversed. A small group of migrants leaves a population (or is accidentally separated from it) and founds a new population in a location where the species is not yet present. Because the sample is random, the set of alleles carried by the migrants may have a very different make up from that of the original population. The population arising from this founding nucleus will thus display a noticeably different genetic structure.[9] Some evolutionists believe this phenomenon is important in speciation.

## The meanders of ancestry

We have just seen that genetic drift exists because populations never reach the infinite size that would allow us to apply the law of large numbers and to describe the exact change in gene frequency in the population. We must accept that allele frequencies fluctuate randomly, and all the more so, the smaller the population. We have mentioned the effects of population bottlenecks and migrant founders on allele diversity. We can also study the effect of a reduction in population size on inbreeding. This is a way of investigating the origin of the human species on the assumption that there was a single ancestral population.

### Genetic drift, inbreeding and gene coalescence

Inbreeding is mating between related individuals. It concerns all genes and increases the number of individuals who are homozygotes in a population. On the other hand, homogamy - mating of individuals with similar phenotypes - gives rise to homozygotes only for those genes involved in the similarity and not for others.

The inbreeding coefficient is the probability that both gene copies at a given locus in an individual are copies of a single ancestral gene. This coefficient is one of the most complex concepts in population genetics. It can be used, for instance, to measure a deficit in heterozygotes due to preferential mating between relatives, the reference being a random mating population with the same allele frequencies and for which the inbreeding coefficient is arbitrarily set at 0.

On the other hand, in a small population, inbreeding occurs even when mating is random. The smaller the population, the faster the increase in the inbreeding coefficient. Of course, the calculated value is only an expected value because of the effects of chance but we always end up with a coefficient of 1. Not only is allele diversity lost but the entire population's genes, at the locus under consideration, end up by being copies of just one of the starting genes. This means that, if we climb high enough up the phylogenetic tree, we can reach the ancestral gene of all current genes for a given locus.[10] This is known as gene coalescence in a population (see annex MA2).

Coalescence concerns all loci. In sexually reproducing organisms, however, the ancestral genes for two distinct loci were usually not present in the same individual at the same time. Moreover, because of within-gene recombinations, adjacent parts of the same gene can derive from different ancestors. Coalescence is a random phenomenon. In the absence of selection, all the genes present at any given time have the same probability of being the

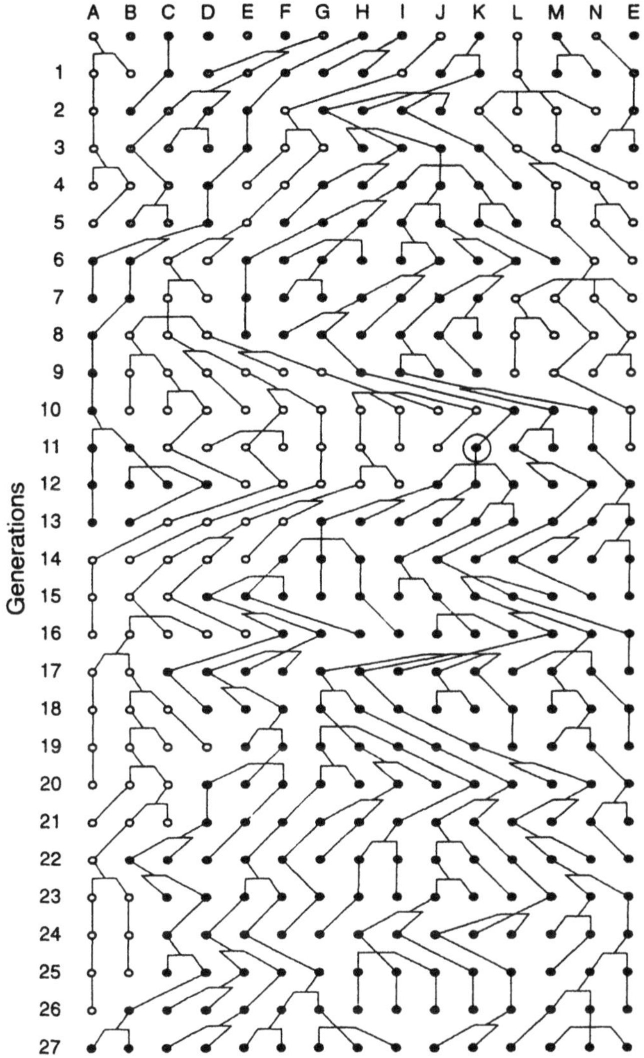


*Figure 3.2.* Simulation of gene coalescence in a haploid population kept at a constant number of 15 individuals. Coalescence obeys the same principles in haploid and diploid species. At a given generation, each gene is the copy of just one ancestral gene of the previous generation (except for recombinations). This is why it is easier to explain coalescence in haploid than diploid species. In this figure, at each generation, each individual is begotten by an individual who is randomly selected among the 15 members of the previous generation who thus have the same probability of reproducing; on average they leave behind one descendant (sometimes none, sometimes several). At the last generation (n° 27), the 15 genes all originate from the same gene (identified by a circle) 16 generations earlier. Of the 15 starting genes, all but one (allele K) have disappeared.

lucky one, by which we mean the future ancestor of all the genes present in the population formed by later generations. We cannot tell how long this process will take. Coalescence time is a random variable for which we can only calculate an expected value. On average, for $k$ genes, we have to go back $2k$ generations to find a single ancestor common to all. For a diploid population of a constant number N of individuals, i.e. for 2N genes situated at the same locus, we can expect to have to go back 4N generations to find the ancestor. Conversely, an average of 4N generations are needed for all the genes to be copies of just one of the 2N genes present at a given time.

**Did mitochondrial Eve meet the Adam of chromosome Y?**

It is easy to calculate the number of ancestors each of us had in Charlemagne's time. We are separated from this emperor with the white flowing beard by 1200 years, that is by about 48 generations of 25 years. In Charlemagne's day, we would thus have had $2^{48}$ (about $10^{14}$) ancestors. Assuming that the human population of the time was $10^8$, we can conclude that each of Charlemagne's contemporaries was on average $10^6$ times our ancestor. Moreover, this is an underestimate because they did not all have offspring. We are all descendants of Charlemagne (if the line did not die out with his sons Charles, Lothaire and Louis and their progeny); we are all related! And what if we look even further back into the past?

The Earth's inhabitants all belong to the species *Homo sapiens sapiens* which, like the now extinct *Homo sapiens neanderthalensis*, was preceded by an archaic species *Homo erectus* which was, in turn, preceded by *Homo habilis*. This human lineage was probably born in Africa, about 5 to 10 million years ago, after diverging from the great African apes as summed up in the title of a book by the French paleontologist Yves Coppens (*"Le Singe, l'Afrique et l'Homme"* - *"The Ape, Africa and Man"*). But when and where did modern man appear? Many anthropologists claim that his earliest fossil remains are African and date back 100 to 200 thousand years but others think that fossil remains found in Asia might be those of our direct ancestors. There is some consensus of opinion on the following scenario. Populations of *Homo erectus* probably left Africa a million years ago and went off to settle in other regions of the Old World. The Man from Java and the Man from Peking would be their descendants. Archaic *Homo sapiens* appeared some 250 000 years ago. It is a "hotch-potch" group in which were included the Neanderthalians from Western Asia and Europe although, in July 1997, geneticists working on fossil DNA claimed that the Man from Neanderthal had had a separate evolutionary fate. At this point, the scenario diverges:

1. In the first multiple region model, the process that had given birth to archaic *sapiens* populations went on to yield modern populations.

Widespread interbreeding between populations established early on a continuity in the anatomic features of populations from different regions. Genetic roots would thus be very deep-seated and run back to the time when populations of *Homo erectus* reached different parts of the planet.

2. In the second model, modern mankind comes from a single population of African archaic *sapiens* whose descendants spread throughout the Old World and gradually replaced all other populations of archaic *sapiens*. According to this model, there is no continuity in regional features. The genetic roots are much less deep and just go back to when each population was founded. This scenario has been called the "Noah's Ark" or "Out of Africa" model and is related to the so-called "mitochondrial Eve" model.

*Figure 3.3.* "African Eve", our common ancestor, has been the subject of the lead article in most of the world's newspapers. Above, the January 1988 issue of Newsweek. © 1988 Newsweek, Inc. All rights reserved. Reprinted by permission.

The mitochondrial Eve model was proposed by Allan Wilson and his colleagues in an article published in January 1987 in *Nature*.[11] This model supported the notion that molecular biology can provide us with further evidence on the origins of the first *Homo sapiens*. Their mother, symbolically called Eve, would have lived in Africa about 200 000 years ago. (Other work suggests that Adam grew up in the land of the Aka pygmies.) Wilson studied the mitochondrial DNA of our contemporaries. DNA is found not only in the cell nucleus which harbours the chromosomes but also in other organelles, especially in the mitochondria which are the cell's "power stations". In mammals, mitochondria are passed on by females only and thus the study of mitochondrial DNA tells us about maternal relationships.

A few years earlier, in 1980, W. Brown had shown that the mitochondrial DNA of 21 men from different geographic areas and of different ethnic origins differed little as if, some 200 or 300 thousand years ago, the number of females with fertile offspring had been considerably reduced. Rebecca Cann, Mark Stoneking and Allan Wilson compared the mitochondrial DNA of 147 individuals in populations from five different geographical areas: Africa (the African sample was composed mainly of American Blacks), Asia, Australia, New Guinea, and Europe. They treated the mitochondrial DNA with 12 different restriction enzymes. Restriction enzymes cleave nucleotide chains at specific points into sequences of a few nucleotides and each enzyme thus provides a DNA-specific restriction map. A comparison of all the maps distinguished 134 different DNA types. Calculations were performed to find out how related they were and highlighted two main branches in the evolution of the mitochondrial DNA of *Homo sapiens*, one comprising only Africans, the other individuals from all five geographical areas. The most logical explanation for this result is that the common ancestor of all mitochondrial DNA from current populations comes from Africa. If we follow the path from DNA to the individual to the family tree, we find "Eve"; she was black, and the Garden of Eden was in Africa. When did she live? We can answer this question by converting the number of observed differences into accrued mutations and years passed. Using this "molecular clock", the authors of the paper estimated that Eve lived between 280 000 and 140 000 years ago. They determined the time constant of the clock using variations in the mitochondrial DNA from 18 tribes in New Guinea assuming that each tribe's DNA originated from a single founding mother. The measured mutation rate was 2 to 4 percent per million years.

Hardly surprisingly these results provoked uproarious debate in the scientific community. From the beginning of the 1960s, the small core of researchers investigating the origins of mankind had been torn between physical and molecular anthropology. Although the African Eve affair led to

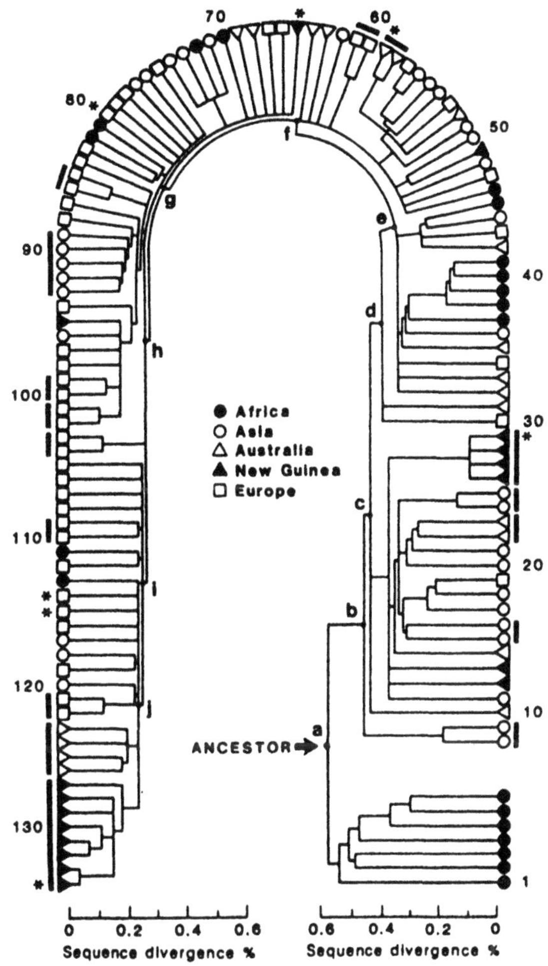


*Figure 3.4.* Allan Wilson's team in California was the first to suggest that molecular biology can take us back to our origins.[11] The study material was the DNA present in mitochondria which is passed on from one generation to the next by women only. It is thus a much better material for investigating kinship than nuclear DNA which is inherited from both parents and which undergoes recombination. By analysing mitochondrial DNA from 147 individuals from five geographic regions using restriction enzymes, Wilson distinguished 134 different DNA types. The differences were used to build phylogenetic trees by the method of parsimony. The tree illustrated in the figure is one of the most parsimonious. Africans are present in both primary branches and, in one of these branches, they are on their own thus supporting the hypothesis that the origins of mankind lie in Africa. All the populations examined except the African population have multiple origins. Reprinted by permission from *Nature*, 325, p. 34, © 1987, Macmillan Magazines, Ltd.

clashes between individuals or schools of thought, it also offered interesting insight and information on the application of molecular biology techniques to the study of life. Three of the issues broached were:

1. A tree should not hide a forest from view. In tracing the evolutionary history of the mitochondrial DNA sequences, Wilson's team used the parsimony method which selects the tree built in the fewest "steps" (or rather with the fewest branches) to derive relationships among sequences. For instance, with 6 DNA types, we can build 105 different trees. Amongst these, we have to choose the most parsimonious without knowing which of the DNAs is the starting DNA (that is, without being able to "root" the tree with any certainty). Wilson's results concerned 134 DNA types. There were thus more than $7 \times 10^{35}$ possible trees of which over 10 000 were considered parsimonious! His choice of tree was, of course, criticised. There were criticisms of how he established the mutation rate and how he implicitly assumed that it was constant in all the tree's branches. There were also criticisms of his choice of geographic area. To these critics his results showed that Eve could in fact have come from virtually any one of the areas and especially from Asia or from both Asia and Africa.

2. A dichotomic scheme based on sequence analysis should not be mistaken for a genealogical tree. If bifurcations are to be turned into trees, we must know in which direction the traits under study evolved and distinguish similarities due to close kinship (sharing derived traits) from similarities inherited from a long-gone common ancestor (sharing primitive traits). Only the former can provide precise information on genealogy. Unlike cladistics, molecular biology cannot discriminate between primitive and derived traits. We could resort to physical anthropology but with no guarantee of success.

3. What should we think of the famous bottleneck apparently undergone by mankind a few hundred thousand years ago? Wilson himself adopted W. Brown's idea and admits that it is tempting to establish a link between the appearance of ancestral type DNA and the drastic decrease in population size (bottleneck).[12] According to this hypothesis, even if Eve were not one of the partners in a unique couple, she would nevertheless be one of the females of a restricted population with an exceptional destiny. But let's not be too hasty. We must not confuse the genealogy of individuals and the genealogy of genes. Moreover, it is difficult to analyse the way genes behave during bottlenecks because of the importance that random fluctuations may have at the time.[13] Simulations nevertheless indicate that a very narrow or long-lasting bottleneck can decrease polymorphism markedly. To account for the fact that there are currently different alleles for a given locus in a species, we have to assume that the original population was at least 500 strong. These estimations are contrary to the hypothesis that species develop from small

populations in which random fluctuations in gene frequency amplify the effects of natural selection. According to Jan Klein and his colleagues, Wilson's study shows at best that all DNA molecules in present-day mitochondria come from an ancestral molecule borne by a woman who lived some 200 000 years ago.[13]

And what about Adam? Who would want to be an Eve without her Adam with his Y chromosome? Wilson's work was criticised for neglecting this chromosome. Even if mitochondrial DNA went through a difficult patch 200 000 years ago, should we link the emergence of modern mankind with this episode? How did the Y chromosome of males evolve? If Eve really is the woman who carried the mitochondrion from which all our mitochondria are descended, and if Adam was the man who carried the Y chromosome from which all our Y chromosomes are descended, then it is highly likely that Adam and Eve lived at different times and in different places!

Eve is thus still looking for her Garden of Eden but the controversies she is kindling on genealogic trees have a fascination of their own. They urge us to heed the criteria we choose in phylogeny studies and to keep on drawing comparisons between genetic data, palaeontologic observations, linguistic analyses, *etc.*[14]

## Toward reconciling Darwinians and Mendelians

As we mentioned in the Introduction, the challenge of the first 30 years of the 20$^{th}$ century was to reconcile Darwinians (who postulate that evolution occurs by natural selection) and Mendelians (who propose mutation). We have seen that we must not choose between these two forces of evolution but must take both into account and even add migration and genetic drift. Major contributions of population genetics have been its global approach to the study of how gene frequencies evolve and its use of mathematical tools. Nevertheless, the results obtained remain highly speculative as the debates on the origins of mankind - which have stressed the usefulness but also limitations of the approach - illustrate. We shall now outline the main models of population genetics.

The first model (published in 1908) of the genotypic composition of a population was devised simultaneously by two scientists working independently, Godfrey H. Hardy (1877-1947) and Wilhelm Weinberg (1862-1937). The law derived from this model bears their joint names. They based the model on the following hypotheses: gametes associate randomly with respect to the genes under study (hypothesis of panmixia or random mating); the population is infinite so that the law of large numbers applies (the frequency of an event is equal to its probability); and gene frequencies are not influenced by any evolutionary pressure (mutation, gene selection, or migration). Using these simplifications, we can show, on the one hand, that genotype frequency can be deduced directly from allele frequency (if p and q are the frequencies of the alleles A and a, then, according to the hypothesis of panmixia, the frequencies of the two homozygote genotypes AA and aa and of the heterozygote genotype Aa are p2, q2, and 2pq, respectively) and, on the other hand, that allele frequencies - and thus genotype and phenotype frequencies - do not vary from one generation to the next. The Hardy-Weinberg law summarises these results as follows: In a closed population of infinite members, not subject to selection and with no mutation, allele frequencies are constant. If mating is panmictic, genotype frequencies can be deduced directly from a knowledge of allele frequencies; they are also constant. In 1908, Hardy wrote:

> *In a word, there is not the slightest foundation for the idea that a dominant character should show a tendency to spread over a whole population, or that a recessive character should tend to die out.*

The Hardy-Weinberg law is the basis of population genetics. It constitutes the "null model" for evolution and simply states that if nothing causes any change, then nothing changes. For some, this law is a truism, others compare it to the principle of inertia in physics. Calculations and observations on the genotype structure of populations most often refer to this law or to one of its variations. When assessing its contribution to our understanding of real populations, we must clearly distinguish its two facets: (i) If nothing arises that might change allele frequencies (a situation of isolation, infinite population size, no mutation, no selection), these frequencies do not change. This may seem trivial but it contradicts several popular misconceptions such as the idea that a dominant gene tends to invade a population, or that allele ratios spontaneously tend toward a 50/50 equality. (ii) In panmixia, genotype ratios are deduced in a simple fashion from allele ratios. This second facet of the law is often borne out in real populations. It must be totally dissociated from the first which is never confirmed but is instructive.

**The Wahlund effect**

We obtain the Hardy-Weinberg frequencies (p2, 2pq, q2) only when the entire population is panmictic. Many populations, however, are made up of groups (or demes) that mate preferentially. The outcome of this is the Wahlund effect. When we combine these panmictic subpopulations with different allele frequencies (in theory not in practice!), we note an overall deficit in heterozygotes relative to the Hardy-Weinberg structure even though this is the structure of each subpopulation.

Because it is not always clear at the outset whether the population under study can be considered as a single panmictic unit, the Wahlund effect is always a possible explanation for a heterozygote deficit. For instance, the sample may have included several subpopulations that are geographically isolated enough to be genetically differentiated and to display different allele frequencies (think of plants or even animals that are not very mobile and that are sampled over a large enough area). Or maybe the sample included two populations that cannot interbreed freely, or even several generations of different genetic composition.

During the two decades following Hardy and Weinberg's work, R. Fisher and J.B.S. Haldane in Great Britain and S. Wright in the United States laid the true foundations of population genetics.

Ronald A. Fisher (1890-1962) belonged to the same school of probabilistic statistics as Galton and Pearson did. He was the man behind the Monte Carlo method, estimation theory and the maximum likelihood method. In 1919 he was invited by John Russel to join the team at the experimental station in Rothamsted where he brought about a real revolution in methodology. Values obtained by measurement were no longer considered for what they were but as representations of a set of possibilities with probabilities attached. Fisher considered extensive and complex experiments far more effective than simple experiments. In agricultural experimentation, he said, no aphorism is repeated more often than the one saying that we can ask nature only a few questions at a time, or even just a single question. To him this was completely mistaken. Nature willingly provides answers to a carefully planned logical questionnaire. Even if we were to ask but a single question, nature would refuse to reply before another question had been resolved.[15]

Because experimental results fluctuate, they have to be analysed by probabilistic methods. We can estimate mean values and deviations, use significance tests, and eliminate or reduce the causes of systematic errors. This methodology in hand, Fisher tried to use the action of selection and mutation to explain several of life's features, such as the evolution of the system of dominance/recessivity or the evolution rate of a phenotypic character.

Sewall Wright (1889-1988) was more interested in the effects of drift than was Fisher. A major contribution of his was the shifting balance theory. Wright analysed the different kinds of evolutionary pressure and concluded that a large population subdivided into subpopulations encourages rapid evolutionary progress. Because of drift, several small populations explore more avenues than does a single large population. If one such small population "finds" a more effective evolutionary path, its offspring will invade the species, thus producing rapid evolution and an apparent "jump". [16]

Although John B.S. Haldane's (1892-1964) approach was more diverse and biological than Fisher's, he also supported the use of mathematics:

> *At present one may say that the mathematical theory of evolution is in a somewhat unfortunate position, too mathematical to interest most biologists, and not sufficiently mathematical to interest most mathematicians. Nevertheless, it is reasonable to suppose that in the next half century it will be developed into a respectable branch of Applied Mathematics.*[17]

The legacy left by Fisher, Wright and Haldane was chiefly theoretical but it was so astonishingly coherent [18] and convincing that many naturalists applied their models to their own studies. From this urge to put population genetics to the test arose ecological genetics.

## 2. HEYDAY AND DECLINE OF ECOLOGICAL GENETICS

### A story about moths

In 1895, sixty years had passed since the expedition of the *Beagle*. The Universal Exhibition in 1851 had marked Great Britain out as the workshop of the world, and the industrial revolution, begun in the 18th century, continued to tap the country's human and other resources. Coal was particularly heavily exploited, and coal means ... pollution. This is when a simple moth entered stage. The peppered moth, or *Biston betularia*, has white wings mottled with black. It holds a very special position in the bestiary of the theories of evolution because it provided a concrete example of natural selection, the sort of example that Darwin's work lacked. Darwin had had to base his theory on observations of how breeders practised artificial selection.

The first capture of a black or melanic form of the peppered moth was recorded in 1848 [19] in the city of Manchester in the heart of England's oldest industrial area. The frequency of this form, baptized *carbonaria*, increased rapidly and, by 1895, 98 percent of the *Biston betularia* specimens captured in Manchester were melanic. The melanic form also turned up in many other areas. Clearly, something was up. In 1900 therefore, the Evolution Committee of the *Royal Society* in London set up an enquiry with the help of its corresponding members. The enquiry concluded that melanism was linked to industrialisation since in rural areas only the speckled white form (called *typica*) was to be found. Naturalists began to speak of industrial melanisation and wondered why the moths darkened as the landscape of industrial England blackened under the effects of soot and other pollutants.

Several scientists offered Lamarckian explanations such as a physiological effect caused by pollution. Thus, in 1926, in line with contemporary mutationist thinking, a geneticist argued that melanism was due to the mutagenic action of certain pollutants and undertook experiments to prove this. On the other hand, Darwinian-type explanations put the emphasis on selection through camouflage. Like many night moths, adult *Biston betularia* rest by day on tree trunks and in places where they are difficult to detect. In 1896, J.W. Tutt suggested that, because pollution darkens the trunks, camouflage might explain melanism. However, it was not until 1950 that the British biologist H.B.D. Kettlewell conducted a remarkable set of experiments which were to turn the industrial melanism of the peppered moth into a case study found in every textbook on evolution. (Industrial melanism was not encountered in the peppered moth only but also in more than 100 other species of moth as well as in other insects and arthropods, in several European countries and in North America.)

Kettlewell was a student of E.B. Ford, the founding father of ecological genetics. Ford was the leader of a school of English naturalists who, under the intellectual influence of Fisher, were out to demonstrate the action of selection in nature. They studied various kinds of polymorphism in animals (especially butterflies, a field in which Ford was an expert) and plants. Ford's work *Ecological genetics* is a compilation of many of these studies. Industrial melanism, the subject of Kettlewell's investigation, comes across as one of its most convincing case studies. By a series of controlled crosses, Kettlewell established the genetics of melanism in the peppered moth. The phenotype is due to a dominant gene denoted C. CC and Cc individuals are black (*carbonaria*), cc recessive homozygotes are light (*typica*). He also drew up a detailed distribution map of *Biston betalaria* forms in Great Britain which confirmed the relationship between the high frequency of melanics and industrialisation. To catch his specimens by night, Kettlewell used mercury vapour lamps after ensuring that they were as effective in attracting the light and dark forms. He counted the number of specimens of each form captured at each site. Eighty-five percent of the specimens were melanic in Birmingham Park whereas no melanic moth was found in Dean End Wood, a rural area in South Wales (see Table 3.1). He then measured mortality at both sites using conventional capture-tagging-release-recapture experiments and found that mortality of the dark form was double that of the light form in Dean End Wood whereas the converse was true in Birmingham. Kettlewell suggested that at each site the least "cryptic" form (the one easiest to detect) was prey to birds. His peers were at first rather sceptical about this selective prey theory. They were only convinced after seeing on site or on film how the moths on tree trunks were preyed upon by various species of sparrow.

*Table 3.1.* Differences in the viability of the two forms of peppered moth in two environments as given by the results of capture-tagging-release-recapture experiments and by direct observation of preying by H.B.D. Kettlewell and his team.

| | Polluted area (Birmingham) | | | Unpolluted area (Dean End Wood) | | |
|---|---|---|---|---|---|---|
| Phenotype | *typica* | *carbonaria* | Total | *typica* | *carbonaria* | Total |
| Captured | 63 (11%) | 528 (89%) | 571 | 297 (100%) | 0 | 297 |
| 1st test (Tagging-release-recapture) | | | | | | |
| • released | 64 | 154 | | 496 | 473 | |
| • recaptured | 16 (25%) | 82 (52%) | | 62 (12%) | 30 (6%) | |
| 2nd test (Predation by birds - the same number of butterflies of each phenotype were exposed) | 43 (74%) | 15 (26%) | 58 | 26 (14%) | 164 (86%) | 190 |

The selective process described by Kettlewell is simple. In rural areas the *typica* form of the peppered moth blended in perfectly with the mottled lichen sheath on tree trunks. These moths thus escaped from predators that hunt by sight. In industrial zones, the lichens disappeared because of pollutants, especially sulphur dioxide. The barks were left bare, became even darker with soot, and thus provided better camouflage for the *carbonaria*.

It was a splendid demonstration of how selection (predation) in nature benefits one form at the expense of another. The very marked selective difference between the two forms meant that the evolutionary change occurred incredibly fast despite there being just one generation of peppered moth per year. It was much faster than most Darwinian naturalists thought possible. Until then, they had believed that evolution by natural selection was only perceptible over quasi-geological time-spans. Demonstrating the importance of selective effects in nature is one of the most significant contributions of ecological genetics. However, this was not quite the end of the story of industrial melanism in England. The 1970s saw a case of inversion of evolution.[20] The frequency of the *carbonaria* form decreased in many regions apparently in line with the reduction in pollution after the Clean Air Act.

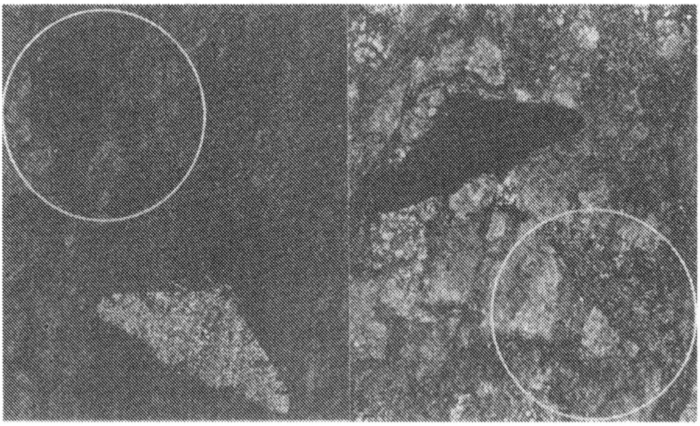

*Figure 3.5.* The *carbonaria* and *typica* forms of the peppered moth on a tree trunk devoid of lichen in a polluted forest (left) and on a trunk in an unpolluted forest (right). In each case, one of the forms is much more visible than the other. The mottling of the *typica* imitates the lichen.

Kettlewell's studies left unsolved problems. Why weren't the *typica* eliminated in polluted areas such as Manchester if their fitness in this environment was really lower? In the years between 1970 and 1980, a team of British researchers undertook a detailed analysis of a 50-kilometer stretch of land between Liverpool (a polluted zone with a high frequency of melanics) and North Wales (a rural zone where *typica* predominated).[21] They conducted experiments similar to Kettlewell's. They measured the frequency of the two forms at several points along the route (Fig. 3.6A) and the extent to which each form was preyed upon (Fig. 3.6B). Figure 3.6A shows that there was a gradual decrease in the frequency of *carbonaria* from Liverpool to Wales. Such graded variation in a character along a transect is called a frequency cline. It is unexpected if just predation were involved. In sites where one of two forms is a prey of choice, the more vulnerable form is eliminated irrespective of whether there is a large or small difference in the extent of predation. (Predation only influences the elimination rate; the frequency curve should look like a flight of stairs (Fig. 3.6C)).

The observed cline could only be explained by taking migration into account. A field study confirmed this hypothesis. The males fly out at night to seek females. The females stay put and secrete pheromones to attract the males. By placing a network of traps at different distances all around the release-point of tagged male moths, the British researchers established that they regularly flew several kilometres each night and that the moths flew off in all directions.

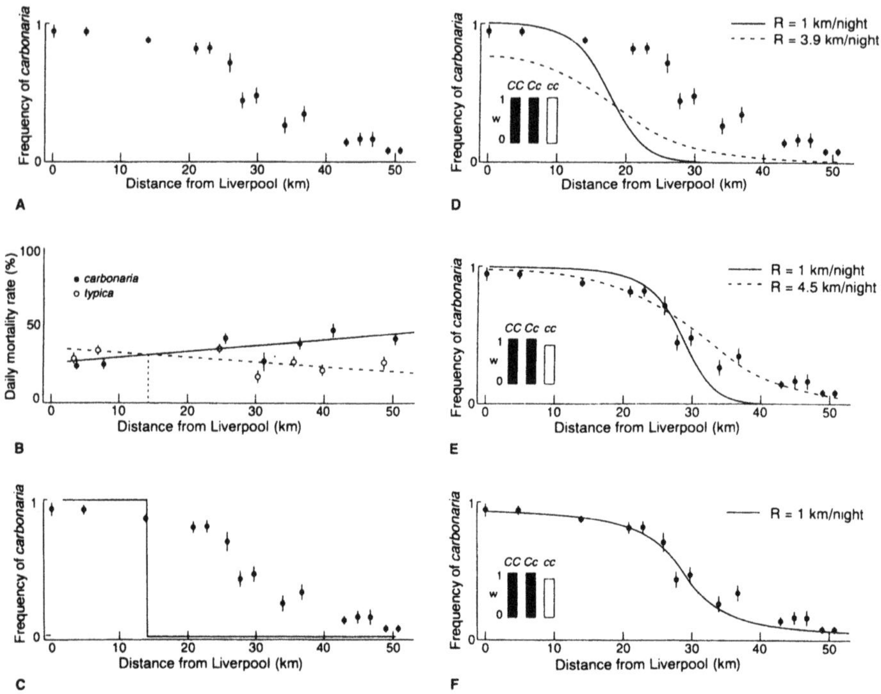


*Figure 3.6.* A. Observed cline (frequencies measured by trapping). The vertical bars represent standard deviations. B. Daily mortality along the cline due to predation. C. Theoretical model for the distribution of the two forms if only predation were involved. D. Model including predation + migration (R= migration by night, in kilometres). E. Model including predation + migration + non-visual selection (relative fitness of *typica* cc = 0.86) F. Model including predation + migration + non-visual selection + frequency-dependent selection.

We can build a theoretical model combining the effects of predation and migration and obtain a family of theoretical clines that are more or less flat depending upon the average distance covered (Fig. 3.6D). However, when we compare these theoretical clines with the observed cline, we note a shift. The theoretical decrease in the frequency of melanics occurs nearly 20 kilometres before the observed cline (as measured from Liverpool). In other words, there is a fairly large area where melanics prevail despite being preyed upon slightly more often. We must therefore postulate a selection process independent of predation that favours melanics as long as they are not at an overwhelming disadvantage when face-to-face with the predator. For the theoretical clines to be centred upon the observed cline, melanics require a calculated selective advantage of about 20 percent over the *typica* (Fig. 3.6E). Data on the viability of various artificially bred forms have confirmed that the melanics are indeed at an advantage, perhaps because they absorb heat better or for some other physiological reason.[22]

Including a second selection process along with predation noticeably improves the fit between theoretical and observed clines. For a migration distance of one kilometre per night, the fit is good over the central part of the cline but not at either end; for a distance of 4 kilometres per night, the reverse is true. To improve the fit, an additional selection process has been postulated; so-called frequency-dependent selection advantages the rarer form at each end of the cline (*typica* in Liverpool, *carbonaria* in Wales). As shown in Figure 3.6F, this leads to theoretical and observed clines with near-perfect fit. Frequency-dependent selection has been observed in quite a few cases, in particular when vertebrate predators are present. It is a plausible explanation but, to our knowledge, has yet to be demonstrated for *Biston betularia*.

This example is a good illustration of the kind of approach that ecological geneticists take to identify the evolutionary factors maintaining polymorphism. It also highlights the limitations of the approach. As soon as we examine things in detail, we discover that many factors intervene in nature, that they interact and are often difficult to put to the test.

## The strategy of ecological genetics

The beginnings of ecological genetics were marked by the analysis of industrial melanism. Ecological genetics springs directly from population genetics and proceeds in five steps:

1. finding a polymorphism
2. establishing the genetic determinism of this polymorphism
3. mapping its distribution
4. deducing correlations between alleles and environment
5. testing these causal relationships experimentally.

Because the first step is seeking polymorphism, ecological genetics has to come to grips with the following questions: What is a population (beyond the empirical concept we have used so far)? What is the rationale for biodiversity? Why are species in the same environment so diverse? Why are individuals within a given species so different? Why do some species need specific environments or vegetation whereas others can survive under much more varied circumstances?

## In raptures over thyme essences

We shall use a typical study in ecological genetics to try to find answers to these questions. We shall describe the steps followed and the obstacles encountered.

First, we have to select an appropriate organism; we shall go for thyme. This species of the *Labiacae* family can be fairly easily defined, from the standpoint of taxonomy, and is found in a wide range of plant combinations in the Mediterranean region. Most importantly, two of its easily observable characteristics display remarkable genetic diversity: sexual form (to be discussed later) and chemical form (thyme essences have different compositions).

Thyme is well known for its culinary and medicinal properties. These are due to monoterpenes produced by cells under the leaves. (Monoterpenes are molecules formed by the coupling of two 5-carbon entities called isoprenes). An expert with a "nose" can identify several chemical forms by their fragrance, either between populations or even within a population, although chromatography is more reliable and practical especially when we are dealing with many samples. The chemical characters are not due to the environment as shown by transplanting specimens into a new site; transplanted specimens pass on their chemical form to the next generation. We can thus ask: Do the chemical forms reflect a specific diversity (in which case, we could use chemotaxonomy for classification purposes) or, on the contrary, is this a case of gene polymorphism within a species?

### Genetic determinism

Each thyme plant usually produces an essence that contains a predominant monoterpene together with its closest precursors (the molecules that make up its biosynthetic pathway). Depending on the genotype of the plant, the monoterpene is either an acyclic compound (geraniol = G or linalol = L), a non-phenolic cyclic compound (α-terpineol = A or thuyanol = U) or a phenol (carvacrol = C or thymol = T). We can thus define six types of plants, or chemotypes, which are named after the main monoterpene in their essence. For example, the thymol (or T) chemotype denotes all plants in which thymol is the main component of its essence. The thymol and carvacrol chemotypes are phenolic chemotypes.

The type of monoterpene in the essence (chemotype) is governed by a series of interacting loci (epistatic loci). The phenotypic expression of one gene masks that of another gene that is not an allele of the first locus. This phenomenon differs from dominance and only concerns relationships among alleles. The scheme for thyme is the following:

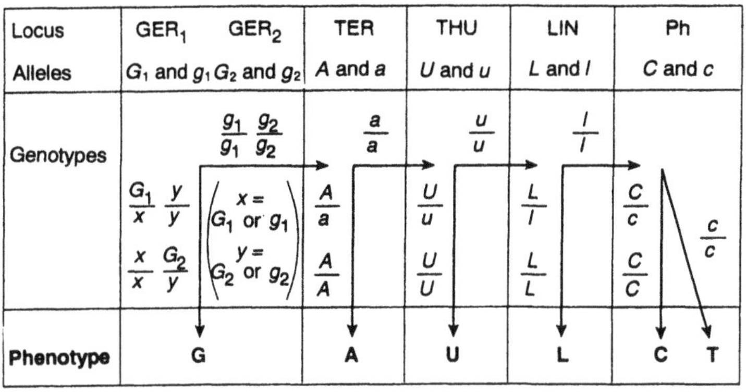


*Figure 3.7.* Genetic determination of chemotype in thyme. This scheme shows the epistatic relationships among loci only and not the biosynthetic pathway of monoterpenes (which is not quite as simple). In each locus, the presence of the dominant allele (in capitals) blocks the biosynthetic pathway at the level of the eponymous monoterpene; only the recessive homozygote genotype determines chemotype at the next locus. Thus, at loci GER1 and GER2, there are eight possible genotypes leading to genotype G, and one genotype (the recessive double homozygote) that puts the decision off to locus TER.

The genetic determinism of the chemotype of thyme displays two features of interest:

1. First feature: the kinds of genes involved. Juxtaposing the scheme of the genetic relationships among loci with the monoterpene biosynthesis pathway established by organic chemists indicates that the dominant genes act by blocking monoterpene production. In some thyme plants, production is blocked several months after germination; for instance, in their early life, L plants can synthesise T and C. From this we can conclude that the genes involved are probably regulatory genes and not structural genes (that is to say, they are probably not the genes coding for the enzymes in monoterpene biosynthesis but for the proteins controlling the expression of these enzymes). This is an important point as most evolutionists now agree that evolution (or, at least, adaptation) is due to changes occurring primarily in regulatory genes. It was also a serious obstacle in that genetic ecologists routinely employed a highly practical method for tracking down different forms of the same enzyme (allozymes) (see description of electrophoresis techniques below) which was however not applicable to the study of the products of regulatory genes. If a regulatory gene acted on the expression of the locus under study, it was customary to abandon the system and declare it unstable.

2. Second feature: the extent of genotypic polymorphism. Plants of the T chemotype, where nothing blocks the pathway leading to thymol, must be of genotypes g1g1, g2g2, aa, uu, ll and cc. They are genetically similar and

homozygotes for the loci under consideration. On the other hand, plants with a non-phenolic chemotype can be of any genotype at the loci that govern the steps following the blocked step; several genotypes correspond to the same phenotype. Consequently, a single genotype specifies the T chemotype whereas 8 x $3^4$, i.e., 648 genotypes specify the G chemotype. (At loci GER1 and GER2, there are 8 genotypes controlling geraniol production. Genotype does not affect phenotype at the four other loci thus leaving three options: dominant homozygote, recessive homozygote, heterozygote). Non-phenolic populations can - and indeed do - harbour a much greater genetic diversity than phenolic populations at these loci.

## Geographic distribution

The study of the distribution of the chemical forms of thyme (and the genes determining these forms) has revealed a transition from the C to T to non-phenolic types when we travel from the Mediterranean toward the Cevennes. The order of the chemotypes encountered along a transect from Montpellier to Vigan is C, T, U, L and A except in a few places such as the basin of Saint-Martin de Londres. This basin is well known for its rather untypical ecological conditions and harbours populations of plants of the non-phenolic chemotype even though it lies at the boundary of zones C and T. To understand this anomaly, we drew a fine scale map of the chemotypes in the basin and surrounding areas. It revealed a highly heterogeneous distribution. The thyme populations on the plateau around the basin were exclusively of the phenolic chemotype whereas those of the basin itself were predominantly non-phenolic. The link between a change in environment and in chemical form could not be due to mere chance. A strong selection pressure is needed to maintain such a clear-cut difference between two populations. On the other hand, the distribution of the C and T populations on the plateau itself, or that of the G, A, U, and L populations in the basin, were more difficult to explain. Neither the effects of chance nor selection could be excluded *a priori*.

Two models can help interpret the above results. The first model, called *stepping stone* model, proposed by Motoo Kimura and Takeo Maruyama in 1971, simulates how a set of adjoining populations - where each population can only exchange genes (or migrants) with its neighbours - evolve under the influence of genetic drift. Each population evolves randomly because of genetic drift but close populations tend to become similar because of migration. If migration is high, all the populations are alike; if migration is low, haphazard zones appear and, after a number of generations, the populations are clearly distinguished with respect to unselected genes (we speak of neutral genes). The simulated map will therefore have a number of clines (or, in our example of thyme, graded series of chemotypes in slightly different proportions). Kimura and Maruyama concluded that when metapopulations occupy large territories they spontaneously split into genetically differentiated subpopulations. This model would seem to provide a satisfactory description of the structure of the vast areas of thyme in the Mediterranean heaths ("garrigues").

*Figure 3.8.* Map of thyme chemotypes near Saint-Martin de Londres. Most thymes can be classified into one of six chemotypes according to the main monoterpene in their essence: T (thymol), C (carvacrol), L (linalol), U (thuyanol), A (a-terpineol), and G (geraniol). In the above map, each population is identified by its two prevailing chemotypes. The basin holds small populations with a variety of chemical forms (white central area) whereas only phenolic chemotypes covering vast areas are present on the surrounding plateau (populations TT, TC, CT, CC). (From the Centre d'études phytosocioloques et écologiques L. Emberger, Montpellier).

*Figure 3.9.* Limit between the plateau and basin of Saint-Martin de Londres. The photograph was taken from the St Loup peak (south eastern corner of the map) in a northerly direction. Because conditions are very different in these two environments (in particular as regards drought), many selective factors could explain the rift between the phenolic phenotypes (on the plateau) and non-phenolic phenotypes (in the basin).

The second model, proposed by Donald Levin in 1978, simulates how a population straddling two environments, in which different alleles are selected, evolves. It can evolve either by remaining polymorphic, if the average migration distance is large enough, or by differentiating into subpopulations, if the distance is small.

These two models, involving neutral genes and selected genes respectively, yield comparable results. Populations can differentiate geographically only if migration is low. Thus, whenever we encounter geographic differentiation in a population, we may assume that the migration distance is small.

We can draw two conclusions from the map of thyme chemotypes around Saint-Martin de Londres. First, a selective pressure keeps the genes that determine non-phenolic chemotypes in the basin and those determining phenolic chemotypes on the plateau. Second, the migration distance is small enough to have allowed the populations confined to the basin to differentiate, either by the action of selection or because of drift, depending upon which of the two models we choose. (In Levin's model, the differences in habitat inducing differential selection have yet to be identified.) These conclusions prompt two questions: What are the selective forces that keep the phenolic and non-phenolic chemotypes so well apart? What is the reason for the low migration, and does it have other consequences?

**The dispersal of pollen and seeds**

We have seen that migration is low enough for the thyme carpeted areas to differentiate. How does this migration occur – by pollen and seeds. Pollen is conveyed by insects only. All studies on pollen dissemination agree that, whereas butterflies transport pollen over considerable distances, bees tend to deposit pollen near the collection point. Most thyme pollen is transported by bees, the remainder by small beetles or ants which transport it scarcely any farther. Under normal circumstances, thyme genes are not carried more than 50 to 100 centimetres by pollen.

Unfortunately, seed migration is not as easy to observe as pollen migration. Its study requires indirect methods and special conditions as those offered by the slopes between the plateau and basin, each with its different chemotype. These slopes are ideal places for measuring migration. Water trickling down the slopes sweeps down the seeds of the plateau population. We would therefore expect to find islands of this chemotype encroaching upon the basin at the foot of ravines. Surprisingly, there is hardly any evidence for this, not because the seeds are deposited along the ravine on the way down, although this does occur, but because selection acts against the genotypes coming from the plateau and in favour of the seeds that travel least and quickly take a hold in the soil. It is apparently difficult to escape from this process of confined migration once it is underway because low migration means that each population adapts increasingly well to its environment and that migrants are less likely to establish themselves. If the environment were perfectly stable, this would be an end to migration but, in reality, new environments are all the time becoming available for colony establishment. For a thyme lineage to colonise them, it must retain the ability to migrate as a seed and, especially, the genetic potential to adapt to a new environment.

**Selective factors**

Thyme plants have to overcome two kinds of constraints, the ones due to the physical environment, the others to competitors and predators. These are fundamentally different. The physical environment does not try to stop the plants from adapting. Thus, if a plant has found a way of growing under high temperatures, the surroundings will tend neither to cool nor to heat up any further. On the other hand, the biotic environment made up of other living things invariably tends to fight any novel ways in which the plants may adapt. If a plant grows to get more light, the shadow it throws over other plants will lead to selection in favour of height increase in these plants.

The plant can probably overcome, more or less, most of the obstacles due to the physical environment. This is not true when the obstacles are other living things. L. Van Valen examined the consequences of such a situation in his Red Queen hypothesis (1973).[23] Engaged in a life struggle, individuals must dip into their resources and invest all they can because evolution will keep only those able to make such a sacrifice. But will this investment not limit the resources available for responding to physiological constraints and induce competition between two kinds of adaptation? In fact, only a species with no competitors or predators could indulge in the luxury of investing enough resources into adapting to the physical environment; cohabiting with other species can only lessen adaptation.

What is the function of the terpenes produced by thyme leaves? Is it still the same as it was when the ancestor of thyme produced the ancestors of modern terpenes? Is it related to the survival of the plant? To answer these questions, we must establish the main causes of thyme mortality.

Various experiments have shown that the renewal rate of thyme is low (under 5 percent per year). If most of the seeds that are produced do not give rise to new individuals, what do they become? In experiments with seed-beds under semi-natural conditions (subject to wind and rain), the seeds stay put and the seedlings appear within 5 centimetres of the point where the seeds were sown. If we choose the right time, we can even spy carpets of seedlings beneath the plants of wild populations. It is after germination that problems occur. On counting the seedlings, we note a drastic decrease in density; some seedlings disappear in a day! Maybe there are predators in the area. The remainder tend to die over the next few days; only a few lucky ones grow beyond the stage of ten leaves (stage at which the samples were taken). Is this due to drought? Or to a microscopic parasite?

Can the herbivores or parasites responsible for the disappearance of the seedlings tell one chemotype from another? To study the "taste buds" of the predators, the seedlings were set before slugs that munch away the leaves one by one at night, with the following results:

1. The slugs had marked gastronomic preferences. They were apparently repelled by some chemotypes (U), barely tasted others (T) and had a liking for (or tolerated) two (A and C).

2. Some slugs at least could distinguish the compounds without tasting the leaves.

3. This selective behaviour was observed only after a training period during which the slugs showed no preferences. This demonstrated their ability to "learn".

Predators thus seem to influence the polymorphism due to terpenes in a way that has yet to be established. However, it is quite unreasonable to wish to model the behaviour of all species that attack thyme. We know of 20 species of snail, at the very least, that go for the plant! We would also have to take into account the presence of plants that compete with thyme. All this would require rather complex experiments.

It is much easier to test the influence of the physical environment. To study how well different chemotypes respond to various physical factors, we used plants of the same origin but of different chemotype. Under restricted lighting conditions, chemotype A plants were more resilient than chemotype T plants, which were, however, advantaged under conditions of drought. To study the effects of high temperatures, we planted different chemotypes in the same pot, keeping humidity high and gradually increasing temperature from 30 to 55°C. Chemotype T displayed the greatest resistance under these conditions.

### Grasp all, lose all

The distribution of the chemical forms of thyme can thus be explained, on the one hand, by the low level of migration and, on the other, by selection due to herbivores and climate. Attempts to combine these selective factors have met with disappointing results because each factor does not act just on thyme plants but also affects the other factors. For example, the grass *Brachypodium ramorum* competes with thyme for light and water. By attracting slugs and snails, it increases the pressure exerted by these predators but, by protecting thyme seedlings from the sun's rays, it prevents drying out. The overall ecological interactions are too complex to predict chemotype distribution. Although the work outlined above has identified selective factors and highlighted an adaptive process, because of the complexity of the situation it falls short of providing an exact description.

In summary, progress in ecological genetics marked time toward the end of the 1970s because of the increasing complexity of the field it addressed. It has revealed the mode of action of various factors on allele frequency but it is uncertain whether it can assist us any further in understanding evolution. As our example of thyme shows, the understanding of gene distribution requires a vast volume of research that ultimately provides rather vague information. One hesitates to undertake such work today. The demonstration of how extensive is gene polymorphism and the controversies surrounding ecological genetics have certainly contributed to its loss of staying power.

## 3. GENE POLYMORPHISM

Life's wealth and complexity have always been a source of astonishment and wonder but also a challenge to those who spend their time contemplating or scrutinising the world. Naturalists like Linnaeus were primarily - one could even say exclusively - concerned with the number of living species and the differences among them. We do not know the exact number as the inventory is far from complete even for groups generally thought to be well known. A new species of palm was discovered in Australia at the end of the 1970s, a new species of monkey in the Gabon in 1984, and three new primates were discovered in the 1990s. There are invertebrates in tropical forests, creatures of the ocean depths and even micro-organisms whose diversity we are only just beginning to fathom. However, despite discovering new species each year, we seem to be faced with ever greater gaps in our inventories. About 1.4 million living species have been listed; how many have yet to be discovered? One million, five, ten, one hundred million? Speculations and extrapolations by experts in systematics abound. Let's note, however, (some consolation!) that current species probably only represent one hundredth of the species that have lived on our planet![24]

Charles Darwin was not the last to be astonished by this biodiversity. In his major work, he provides an explanation:

> *It is interesting to contemplate a tangled bank, clothed with many plants of many kinds, with birds singing on the bushes, with various insects flitting about, and with worms crawling through the damp earth, and to reflect that these elaborately constructed forms, so different from each other, and dependent upon each other in so complex a manner, have all been produced by laws acting around us. These laws, taken in the largest sense, being Growth with Reproduction; Inheritance which is almost implied by reproduction; Variability from the indirect and direct action of the conditions of life and from use and disuse: a Ratio of Increase so high as to lead to a Struggle for Life, and as a consequence to Natural Selection, entailing Divergence of Character and the Extinction of less-improved forms.*[25]

One of Darwin's main arguments to explain evolution by selection was that there is a considerable and apparently endless variability within species that is reflected in often very slight individual differences. At the time, Darwin just noted this variability, but the discoveries of Mendel and of his followers would explain it, and it would become one of the key concepts of population genetics. In 1940 E.B. Ford defined polymorphism as the co-existence within a population of two or more discontinuous forms, in proportions such that the rarer form cannot be maintained just by recurrent mutation.

## Gene polymorphism and how it evolves

The variability in phenotype we have just discussed is due to an underlying polymorphism at the gene level. Gene polymorphism is the co-existence of several alleles at a locus. A locus for which only a single allele is known is monomorphous.

### Cryptopolymorphism

As we have seen, mutation is constantly producing new alleles, in most cases with non beneficial consequences as the mutated gene codes for a protein that functions less well or does not function at all (whether a single amino acid is modified or the protein is truncated as in deletions or nonsense mutations). Sometimes no protein is produced at all. If the carrier of the mutation does not reproduce well or disappears, the allele is said to be sublethal or lethal. Individuals, whether non-viable or sterile, who express a lethal allele (in a phenotypic sense, an allele producing no protein is expressed if this lack of protein has an effect) have zero fitness. Such lethal alleles are most often recessive, sometimes dominant.

Lethal alleles are maintained at certain frequencies in populations. A balance is established between the mutation, which tends to increase the frequency of the lethal allele, and selection, which tends to eliminate the allele. However, because the mutation rate is low (it is often estimated at $10^{-6}$), the frequency of the mutated form is also low and little diversity is engendered by the process. This situation, called cryptopolymorphism, is encountered in nearly all serious genetic diseases such as cystic fibrosis, haemophilia or galactosemia (see mathematical appendix MA3).

### Is polymorphism widespread?

The task of quantifying variations and measuring evolution rates in natural populations has taken much longer than expected but the result has been worth waiting for. Natural populations display a much greater genetic variability, and also evolution rate, than the first evolutionists believed. In 1930, Ronald Fisher demonstrated mathematically that the greater the genetic variation in a population, the more effectively and rapidly selection leads to evolution.

The biologist Francisco J. Ayala confirmed this relationship experimentally using two kinds of fruit-fly populations. The one kind (S for simple) was made up of flies taken from a single site, the other (M for mixed) was obtained by interbreeding flies from two different sites. The M populations had greater genetic variability because they incorporated the pre-existing variants of two populations. Each population was kept in isolation in a "population cage" and provided with constant amounts of food at fixed intervals. The competition between the flies for food and space was fierce because resources were limited; the flies had to evolve rapidly. The experiments were stopped after a given number of generations. The result was as expected. Both S and M populations adapted gradually to the conditions of the population cage, but the M population adapted faster and more successfully because of its initially higher gene polymorphism. There is thus a correlation between variation level and evolution rate. All that remained to be evaluated was the hereditary capacity of the natural populations to evolve. Two models were proposed, one by Muller, the other by Dobzhansky and Ford.

Hermann J. Muller (1890-1967), Nobel prize-winner in physiology and medicine in 1946, was an early collaborator of Thomas Morgan, the father of the chromosome theory of inheritance. In his model, Muller postulated that a population's gene pool comprises almost exclusively "natural (wild) type" alleles that are present at all loci with a frequency of nearly 1. Deleterious alleles arising by mutation and present at a low frequency would make up the rest. (The mutations observed in the experiments by Morgan's team were often deleterious.) According to the model, an individual's normal genotype is that of homozygote with respect to the greatest possible number of alleles; the emergence of heterozygotes is harmful. However, if a mutation were to benefit its carriers, it would spread rapidly through the population and end up by replacing the allele initially considered to be the wild-type allele. In this model, sometimes called the classic model, the magnitude of the hereditary component of variation is low.

In the model of Theodosius Dobzhansky (1900-1975) and Edmund B. Ford (1901-1988), there is no single wild or normal type allele and no variant that is *a priori* any better than any other at all loci. Although some gene variants may be more frequent than others in a population, this may not be so in other populations of the same species. In humans, gene I determines the blood groups A, B and O. This gene has three alleles, Ia, IB and i that have frequencies that vary with population. The frequencies recorded in the French population are 28.7 percent, 5.5 percent, and 65.8 percent, respectively. In all human populations taken together, the frequency of i, which is usually the predominant allele, lies between 45 and 100 percent (100 percent in American Indian populations where all individuals are homozygotes), that of IA between 0 and 55 percent, and that of IB between 0 and 35 percent. In this model, therefore, genetic diversity within populations is the rule and is maintained by various forms of natural selection. Moreover, because genes do not act in isolation, the selective advantage conferred by an allele depends on which other alleles are present in the genotype, besides depending upon environmental factors. (This does not exclude mutations that are harmful to the carrier and are kept at a low frequency by natural selection). In plain words, the ideal type does not exist. In this model, sometimes called the balance model, the emphasis is on inherited variations.

Clearly, these two models reflect contrasting views of the world and of mankind. The classic model, which claims that there are no marked differences among individuals of a population, has led to the idea that races within a species are characterised by great genetic differences. There exists a normal, perfect, homozygote type with respect to which all other types are but inferior and corrupt. Muller was in fact an ardent partisan of eugenics which he took to be the only hope for "*the ultimate genetic improvement of man*". In 1939, he published "The Geneticists' Manifesto"[26] in defence of eugenics which was signed by 22 American and British scientists. After the Second World War, he stuck to his guns whereas most biologists were horrified by Nazi barbarity and repudiated their ideology. It is this implicit notion of a superior race that Dobhansky targeted in 1955 with the balance model. For him, there was no ideal genetic type; only genetic diversity counted.

The discord between these two models (which at times reflected ideological dissension or personal rivalry) was not settled definitively until the second half of the1960s. Until then, conventional genetic techniques had restricted the study of genes to the offspring from the mating of individuals with different forms of a given character. The proportion of offspring of each class indicated whether the differences were of genetic origin, and whether one or more genes were involved. We could study only genes whose allele variety was reflected in observable phenotypic variants but neither invariant genes nor genes that vary but have no phenotypic effect.

## The selectionist versus neutralist hypothesis: Biochemistry and molecular biology shift the ground

Two advances in molecular biology helped overcome the obstacle. The first was a conceptual advance. According to a dogma central to molecular biology, any variation in the amino acid sequence of a protein reflects a variation in the gene encoding the protein. The study of genes can, therefore, be reduced to the study of the amino acid sequences of the corresponding proteins. A gene with no allelic forms codes for an identical protein in all individuals of the population. Conversely, if a protein's sequence differs among individuals, the gene that codes for this protein has several allelic forms.

The second advance was technical and was the development of protein electrophoresis on starch or acrylamide gels. In 1966, H. Harris applied this technique to human proteins and R.C. Lewontin and J.L. Hubby to the proteins of the fruit-fly.

**Enzyme electrophoresis**

Gel electrophoresis evaluates the genetic variability of natural populations by detecting variant proteins. A sample of tissue is taken from each organism. These samples are homogenized separately to release enzymes and other proteins. The supernatants of the homogenates (i.e. the liquid fractions) are then deposited on a starch, agarose or polyacrylamide gel through which a current is passed for several hours. Each protein migrates through the gel in a direction and at a velocity which depends upon its electric charge and molecular size. After a set time, the protein reaches a distance from the migration front that characterises it. The electric current is switched off and the entire gel is covered by a solution containing the enzyme's specific substrate and a dye. The enzyme in the gel catalyses the conversion of the substrate into a product that reacts with the dye, and this reaction gives rise to a coloured band at the protein migration point. Because the different allelic forms of one and the same protein differ in molecular structure and sometimes charge (and therefore have different mobilities in an electric field), the number and position of the coloured bands of the gel highlight, for each individual, the genetic composition of the locus or loci which code for this enzyme.

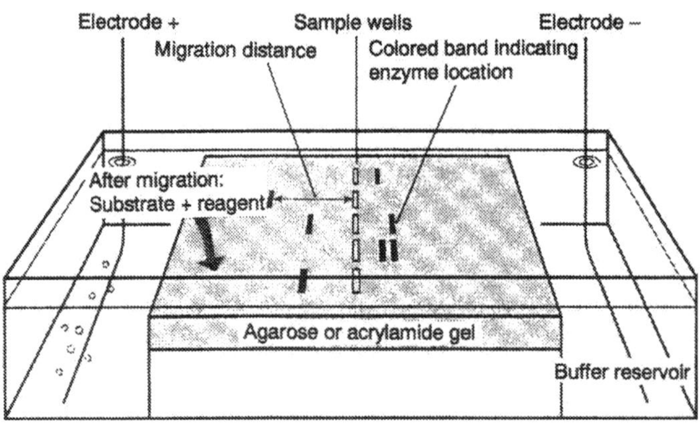

*Figure 3.11.* Electrophoresis gel of enzymes (leucine-aminopeptidases of the cress *Arabidopsis thaliana*). The above gel (shown in part) highlights enzyme variability which reflects underlying genetic variability. Two loci (Lap-1 and Lap-2) are associated with the leucine-aminopeptidase function and thus code for isozymes (enzymes with similar functions). Moreover, Lap-1 has two alleles coding for two allozymes which correspond to the two top bands. (All three possible genotypes are present; the two homozygote types have a single band, the heterozygotes have two). Lap-2 has only one allele; the bottom band corresponds to the enzyme it encodes.

Electrophoresis distinguishes among isozymes, i.e. enzymes encoded by different genes but which display similar functions because they act on the same artificial substrate. Whenever a gene undergoes a mutation, the enzyme it codes for can lose all activity. When this happens, the corresponding band of the gel disappears; we have a nil allele. In actual fact, there are very few nil alleles because they tend to be counterselected and thus persist in populations at very low rates. On the other hand, the mutation is not counterselected when it preserves enzyme activity, at least qualitatively. It then gives rise to an enzyme that is an "allele" of the original enzyme and that is called an allozyme. This occurs in some point mutations where one amino acid is substituted for another. Examination of the genetic code has revealed that the random substitution of a base in the DNA sequence of a gene leads about one time in three to an amino acid being replaced by one with a different charge in the corresponding protein sequence. This alters the overall charge of the protein and, consequently, its electrophoretic mobility. If the change does not affect the enzyme's function, the band will not be suppressed but shifted slightly. By comparing the different runs of a gel, we can thus visualise the diversity of the sample. This kind of protein analysis performed on a sufficient number of individuals often reveals several alleles at a given locus.

More recently, newly developed molecular biology techniques have enabled the study of polymorphism at the DNA level. The technique of DNA sequencing, i.e. determining the order of bases (nucleotides) in the chain, is too cumbersome for the routine study of populations but alternative techniques are available such as the determination of polymorphism at DNA restriction sites. This method has revealed that the number and position of various restriction sites among individuals is highly variable. We speak of restriction fragment length polymorphism (RFLP). It is encountered in all of the genome and especially in its non-coding regions which, because they are not translated into protein, cannot be studied by electrophoresis. Since then, methods to detect polymorphism have flourished (e.g., random amplified polymorphic DNA (RAPD), amplified fragment length polymorphism (AFLP)).

These methods have helped define gene polymorphism more precisely and have resolved the quarrel between Muller and Dobzhansky in Dobzhansky's favour. To detect a polymorphism in a sample of reasonable size, the alleles must have frequencies of the same order of magnitude; no allele should dominate too overtly. In practice, geneticists have set a threshold for polymorphism. Genetic polymorphism is the co-existence within the same population of two or more alleles at a locus, the most frequent having a frequency not exceeding 0.99. The 0.99 threshold is arbitrary; sometimes we choose a 0.95 threshold. The threshold criteria are selected so that they are compatible with the study of a reasonable number of individuals (about 100).

This definition of gene polymorphism however introduces a paradox. If evolution is really driven by selection by the environment, ultimately only the hereditary structure that reproduces most efficiently should survive, and a population should generally have just one allele at each locus, give or take a mutation. Two explanations have been put forward for this paradox. Either a selective mechanism is responsible for maintaining polymorphism (the selectionist hypothesis) or, in addition to the low frequency lethal genes, there is polymorphism among neutral alleles (with no difference in fitness) that is maintained by a balance between mutation and drift. Mutation introduces new alleles, drift eliminates them. This is the neutralist hypothesis. There is probably some truth in both these hypotheses.

## The selectionist hypothesis

According to the selectionist hypothesis, there are several kinds of polymorphism: polymorphism kept in balance by the overdominance of fitness, polymorphism maintained by variability in fitness, or even transient polymorphism.

### Overdominance and sickle-cell anaemia

We speak of overdominance when heterozygotes have a higher degree of fitness than homozygotes. In such cases, polymorphism is stable.

A particularly well informed example of overdominance is sickle-cell anaemia or drepanocytosis. This hereditary disease of the blood which is widespread among the black human populations of Africa, or originating from Africa, is due to an abnormal form of haemoglobin. Human haemoglobin is a protein consisting of four subunits (two α chains and two β chains). The β chain has 148 amino acids with glutamic acid in the sixth position of the chain. A point mutation of the sixth codon of the β gene can convert this glutamic acid into a valine. Haemoglobin with this mutated β chain acquires aberrant properties which are reflected in blood abnormalities in homozygote carriers. The red cells which contain the mutated haemoglobin look like a sickle hence their name. They block the capillaries and lead to fatal anaemia in children. The non-mutated form of the protein is called haemoglobin A, the mutated form haemoglobin S.

The gene of sickle-cell anaemia is also encountered in European populations but at a very low frequency. It is a recessive allele, that is to say the parents of an anaemic child are heterozygotes (their red cells contain the A and S forms) and they do not suffer from anaemia themselves. On the other hand, the homozygote state of the allele is lethal even in children. How can we then explain that the allele is common in populations originating from Africa?

As early as in 1946, Beet noted that haemoglobin S protects against malaria which is rife in the same regions as sickle-cell anaemia. Subsequent research confirmed that, whereas homozygote haemoglobin A carriers suffer and die from malaria, AS heterozygotes - who are not anaemic - are protected against malaria. Endemic malaria thus favours maintenance of the mutated allele, in particular in a zone called the sickle belt which stretches across part of West Africa, Equatorial Africa, Madagascar, Ceylon, and South India. In certain Baamba tribes living near the Ruwenzori Mountains and in the Semliki plain, the proportion of heterozygotes reaches 45 percent.

This polymorphism thus arises from a situation of overdominance where heterozygotes are advantaged. We now need to know what confers them greater fitness in the sickle belt despite the mutated allele being lethal in the homozygote state. The answer is that the red cells of the heterozygotes, which contain a mixture of both haemoglobins, are destroyed faster than normal red cells. This prevents the malaria parasite from completing its development cycle. Heterozygotes, who do not succumb to malaria, have greater longevity and a higher reproduction rate than homozygotes who express normal haemoglobin.

Except in rare instances such as sickle-cell anaemia, the role of overdominance and the action of selection on enzyme polymorphisms remain controversial. We know of many cases where the offspring of mating between individuals of different stock are superior to their parents (taller, faster growth, greater fitness, *etc.*), but we tend to view them as the outcome of the combined beneficial effects of genes inherited from both parents. As long as we have no reason to believe that the parental genes are allelic and that there is no specific advantage in being heterozygote, we can explain this phenomenon of vigorous hybrids, or heterosis, without resorting to overdominance.[27]

Besides, it is often difficult to show that natural selection is acting upon a particular locus, first of all for practical reasons, because the true function of the enzymes detected by electrophoresis is unknown *in vivo*. Thus, the fitness of the different protein variants cannot be explained in many cases. Moreover, testing all possible selective factors is an endless task (see the example of thyme). Finally, the differences in fitness among individuals are often small, and it is a colossal undertaking to obtain data on genotype survival and fertility in different environments. There is an additional theoretical difficulty. Individuals differ in many of their genes and we may never be sure that an observed selective difference is due to the allele under study rather than to alleles at other loci that are statistically correlated with this allele. (This often happens for genetically linked loci).

Because overdominance is rare and enzyme polymorphism is as great in haploid as in diploid organisms (the highest polymorphism levels are observed in bacteria), it is doubtful how great an advantage is heterozygocy in maintaining variability. Overdominance is still a moot point among geneticists, and we have had to seek other explanations for polymorphism.

**Variable fitness**

In a population with constant fitness for different alleles, only overdominance can account for the maintenance of polymorphism at equilibrium. On the other hand, polymorphism can be maintained selectively without overdominance if the fitnesses vary either as a function of gene frequency or in space and over time. We speak of frequency-dependent selection when a genotype's fitness varies with its frequency (and/or with that of other genotypes) in the population. Let us take the example of an allele that is advantaged when rare. Because selection opposes its disappearance, polymorphism is selectively maintained. This kind of selection is called "the advantage of the rare" or "apostatic selection" to remind us that the "unorthodox" allele is advantaged. It applies to all sexual systems insofar as one sex needs the other to reproduce. In other words, the rarer a sex, the more it is sought, and frequency-dependent selection ensures that it is maintained. It also applies to prey-predator systems. Predators tend to attack prey they recognise, that is the most common type of prey, which is thus counterselected until it becomes less common. (We discussed this mechanism when referring to the melanism of the peppered moth.) Because pathogenic organisms evolve by developing virulence against the most common genotypes in host populations, the rare genotypes are advantaged in these populations. Similarly, in a population where different types exploit environmental resources in different ways or have different needs, the rarer types are advantaged because they suffer less from competition from their congeners. They thus acquire greater fitness and will increase in number until they are no longer the rarer type.

Fitnesses can also vary over space and with time. A population's environment is usually not homogeneous but comprises ecological niches, that is micro-habitats that are each best suited to a particular genotype. [28] This heterogeneity (or mosaic as it is sometimes called) helps maintain gene polymorphism but under certain conditions only. Let us imagine that the spatial variations have a "fine mesh", in other words that the mean distance between one habitat and the next is small compared to the distances covered by the individuals, so that during a life-span each individual can try out all the micro-habitats. In this case, the genotype with the best mean fitness over the entire environment has the advantage. It may be the heterozygote which brings us back to the case of overdominance (denoted here as marginal overdominance) but, if it is the homozygote, polymorphism is not maintained. On the other hand, if the environment has a "coarse mesh", each individual will generally only try out just one habitat in its lifetime and, under certain conditions, polymorphism will be maintained even without any kind of overdominance.

One of the best studied cases of polymorphism in a mosaic environment is that of Californian populations of wild oats, an autogamous plant introduced into California by Spanish settlers. In some of these oat populations, genetically different individuals cohabit just a few decimetres apart. Some are adapted to relatively dry and others to more humid micro-habitats. Distribution depends on the micro-relief of the terrain and the nature of the soil. Whenever the micro-habitat changes, one type takes over from the other.

Similarly, changing environmental conditions can encourage different phenotypes alternately and thus maintain polymorphism. These changes must be slow enough for each individual not to have to try out all sets of conditions but fast enough for no allele to become fixed when advantaged. For example, seasonal variations cannot maintain polymorphism in oaks that live for centuries but they do in fruit-flies and in many micro-organisms which present several generations per year.

The way polymorphism is maintained by variation in fitnesses is very similar to the mechanisms invoked to explain the co-existence of different species in the same environment. Just as maintenance of genotype diversity has been attributed to differences in micro-habitats, ecology attributes the maintenance of species diversity to differences in niches. More recently, it has been shown that stochastic processes could lead to the maintenance of the diversity of plant species under variable environmental conditions. Only overdominance is applicable exclusively to intraspecific variability.

**Transient polymorphism**

If a new allele introduced into a population by mutation or migration confers greater fitness to its carriers, it will tend to replace the previous allele, unless it is lost by random drift when still very rare. A gene that has appeared by mutation is present initially only as a single copy in a single individual and, even if it does confer its carrier greater fitness, the likelihood of it being lost during the first generations is high. However, as soon as its frequency has increased substantially, the probability of loss decreases considerably unless the population is really tiny. The substitution of the previous allele by the new one can take some time, especially if the difference in fitness between the carriers of the previous allele and of the new allele is small. During this period, we can observe transient polymorphism.

The industrial melanism of the peppered moth illustrates another kind of transient polymorphism. An allele, that until then had been unfavourable, becomes advantageous because of a change in the environment. The speed at which the substitution took place in this species, which has only one generation per year, is a sign of strong selection exerted by predatory birds.

However, all in all, reversal of the direction in which selection acts and emergence of new advantageous mutants are rare events which doubtless account for very little polymorphism.

**Genetic load**

Because sickle-cell anaemia illustrated overdominance so well (too well), most geneticists considered overdominance a satisfactory explanation for polymorphism for quite some time. Electrophoresis changed all this because, as we have seen, it revealed that polymorphism was too extensive to be explained by overdominance alone. Several arguments then shed doubt on the predominant role of selection in evolution and in maintaining enzyme polymorphism. One such argument related to genetic load.

The word load (*fardeau* in French) should not be construed too hastily in a negative sense. In the 12th century, the arab word *farda* denoted a ball of cloth or merchandise that was carried or placed on a camel's back. In the 17th century, by way of Aragon and Gascony, *farde* (fr) became *hardes* (fr) meaning togs just good enough for beggars on the road. The term *fardeau* acquired its modern accepted meaning from writers such as Jean-Jacques Rousseau for whom "*the weight of an early mistake is a load* (fardeau) *that has to be carried throughout one's life*" or Lamartine who writes in one of his poetic meditations "*to set down the load* (fardeau) *of human misery, is this not dying?*"

> *Thus, death, misery, burdens, errors in one's ways and their punishment, and unimpeachable fate are the ingredients seized upon by the first researchers into heredity in the 19th century. The expression "genetic load" will persist both as a nebulous myth and an inexorable reality related to an unatonable sin.*[29]

The notion of genetic load was introduced in 1950 by Muller to describe the accumulation of unfavourable mutations in a population. The word load in this context thus denotes deficient adaptation as underlined by Haldane in 1937 in relation to deleterious mutations. Both the notion and the word were borrowed in 1966 by R. Lewontin, professor of genetics at Harvard, and J.L. Hubby, when they published the results of their electrophoresis work on fruit-fly polymorphism. They saw this polymorphism, unsuspected until then, as an intolerable burden on populations. The presence of individuals with different fitnesses in a population decreases the overall fitness with respect to the fitness of a population in which all individuals would have the best genotype.

Mutation, which unrelentingly produces "corrupt" individuals, is responsible for a mutation load. During the substitution period, transient polymorphism creates a substitution load. When polymorphism is balanced, maintaining polymorphism in the face of selection produces a genetic load, the segregation load. It is thus named because, at each generation, Mendelian segregation produces individuals of lesser fitness by causing the least advantaged genotype to reappear.[30] Paradoxically, the load in this case is maintained by selection itself (see mathematical appendix MA4).

Irrespective of the kind of selection, the more genes we wish to select at any time, the greater the genetic load (the greater the number of individuals of lesser fitness that cannot reproduce) and, therefore, the smaller the number of begetters or the effective population number, Ne. The result is increased drift of the other genes and a decrease in heterozygotes. Plant or animal breeders are well aware of this problem and take care not to select simultaneously all the traits they desire in a new variety or new race for fear of losing, through drift, many genes they cannot influence directly but that may prove of interest later.

# 4. NEUTRAL THEORY AND THE MOLECULAR CLOCK OF EVOLUTION

## The neutral theory of molecular evolution

To solve the issue of genetic load, we can postulate that the considerable polymorphism of natural populations is not accompanied by any difference in fitnesses. Such a polymorphism is said to be selectively neutral and has no cost.

Another way of coming to the same conclusion is to note that, if the extent of selection acting on different loci varies with locus, preferential reproduction by carriers of beneficial alleles at the loci that are most often selected reduces the effective population number so that selection becomes negligible compared to drift at other loci. These alleles then behave as if they were neutral and we speak of neutral or nearly neutral alleles.

Motoo Kimura, one of the founders of the neutral theory, summarises it as follows:

> *The proposal of the neutral theory (or more precisely the neutral mutation-random drift hypothesis) represented my attempt to answer these questions using the stochastic theory of population genetics. Unlike the traditional synthetic theory (or the neo-Darwinian view), the neutral theory claims that the great majority of evolutionary mutant substitutions are not caused by positive Darwinian selection but by random fixation of selectively neutral or nearly neutral mutants. The theory also asserts that much of the intraspecific genetic variability at the molecular level, such as is manifested in the form of protein polymorphism, is selectively neutral or nearly so, and maintained in the species by the balance between mutational input and random extinction or fixation of alleles. From the standpoint of the neutral theory, evolutionary mutant substitutions and molecular polymorphisms are not two independent phenomena, but simply two aspects of a single phenomenon. In other words, protein polymorphism merely represents a transient phase of molecular evolution.*[31]

Kimura's view was opposed to neo-Darwinian theory only for a time and has now been totally integrated into the theory. The very title of Kimura's book *The Neutral Theory of Molecular Evolution* suggests that Kimura is not proposing an alternative theory of evolution but rather an analysis of Darwinian theory from a molecular standpoint. The main thrust of Kimura's work has been to change the null hypothesis (*a priori* hypothesis) in evolutionary studies. Instead of assuming that all genes are submitted to selection, evolutionists now consider that alleles are neutral unless they have proof of selection.

According to Kimura, the genetic diversity we observe is due simply to the emergence of new neutral forms (or alleles) which in no way modify the carrier. Drift acts on the polymorphism due to these mutations either by eliminating the new allele (this is what happens in most cases) or by fixing it (by eliminating the previous allele). The polymorphism is then transient, lasting only whilst the substitution takes place. To account for it, we must suppose that the populations are large enough for the random fixation of a gene or the loss of the other genes to occur very slowly, and that mutation continually introduces new neutral alleles. The substitution rate of a neutral allele by another during evolution is independent of population size. However, the mean number of alleles at this locus, at any given time, increases with population size and neutral mutation rate (see mathematical appendix MA5).

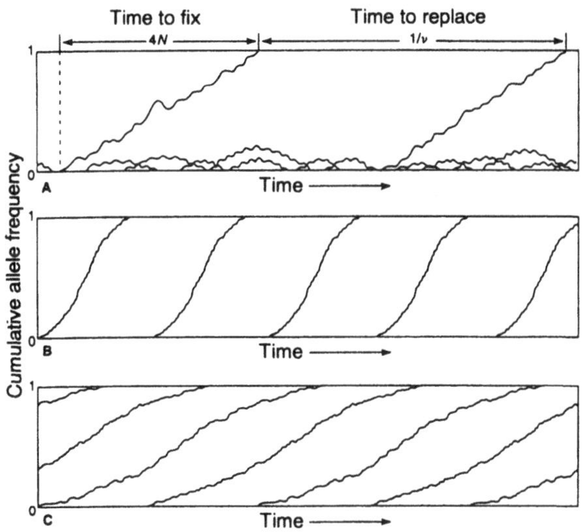


*Figure 3.12.* Alleles are said to be neutral when they do not modify the carrier in any way and are therefore not subject to selection, or when the selection undergone is negligible compared to drift. New neutral alleles continually appear by mutation (A). Most are lost but some end up by establishing themselves; they replace previous alleles. This happens on average every 1/v generation, v being the rate of neutral mutation at the locus under study. When a new neutral allele establishes itself, the time elapsed since it first appeared is on average 4N generations, N being the size of the population. In a small population (B) and in a large population (C), we find the same rate of neutral mutation v and the same rate of allele substitution (equal to the preceding rate since the mean substitution time is 1/v). However, because the time taken by the mutation to establish itself is proportional to N, more neutral polymorphism is encountered at all times in a large than small population.

Neutral mutations are a fraction of the total number of mutations. Mutations can in fact be neutral, lethal or sublethal, or even positively selected. If m is the total number of mutations, $\nu$ the number of neutral mutations, $\mu$ of lethal mutations and $\varepsilon$ of mutations that can be selected, then $m = \nu + \mu + \varepsilon$. The neutralist hypothesis postulates that positively selected mutations are by far the least important category and that the study of genome variations focuses on neutral and lethal mutations. However, because there are few lethal mutations, they tend to be eliminated by selection, and neutral mutations are the most readily observed mutations.

## The molecular clock

The first principle of Kimura's neutral theory is as follows:

*For each protein, the rate of evolution in terms of amino acid substitutions is approximately constant per year per site for various lines, as long as the function and tertiary structure of the molecule remain essentially unaltered.*[31]

This constant rate hypothesis is challenged by many molecular biologists studying evolution. Kimura has written at length on its relevance. We shall discuss its application to phylogenetic tree construction.

It is generally agreed that organisms with a recent common ancestor are more alike than organisms with a common ancestor from a more distant past. This simple hypothesis is the basic logic underlying comparative studies of phylogenetic tree construction, i.e., the evolutionary history of species. By observing similarities among species, we can establish the points at which they successively diverged. For example, let us consider three species A, B, and C. If A and B are more alike than either is like C, their phylogenetic relationship is probably the following:

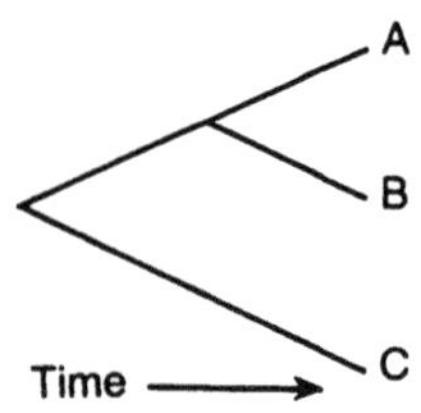


Phylogenetic tree construction is not an easy task. Evolutionary rates can vary considerably depending upon the time period, groups of organisms, or characters of an organism. Moreover, we have to distinguish similarities due to a common ancestor from those due to similar lifestyles or accidental convergence. Identifying characters that can be used to infer phylogeny is one of the hurdles of systematics.

Traditionally, comparative anatomy provides most of the information for establishing a phylogeny of living creatures. Additional data come from embryology, cytology, ethology, biogeography, *etc*, and recently from the study of macromolecules (i.e. proteins and nucleic acids). By comparing macromolecules and analysing their differences we can establish molecular phylogenies. Changes in evolution rates among branches and over time can make things difficult. The molecular approach, however, has two advantages over conventional methods. First of all, it is easy to quantify the information. Once we know the sequence of a given macromolecule in different organisms, we can establish the number of positions that differ in the sequence of each. Because of life's deep-seated biochemical unity, we can thus compare very different types of organisms.[32] Second, the study of these macromolecules provides information on the amount of genetic change undergone. Let A and B be two living species that have evolved from a common ancestral species X which is now extinct. Let v be the rate of allele substitution, that is the rate at which an amino acid has replaced another in the protein sequence under study during evolution. (As we have seen, this rate corresponds to the number of neutral mutations per generation.) Because this rate does not depend upon the environment, it is generally agreed that v has the same value in the two phylogenetic branches, from X to A and from X to B, and that these branches have undergone the same number of substitutions. Consequently, if A and B differ by k amino acids, we assume that there have been k/2 substitutions in each branch.

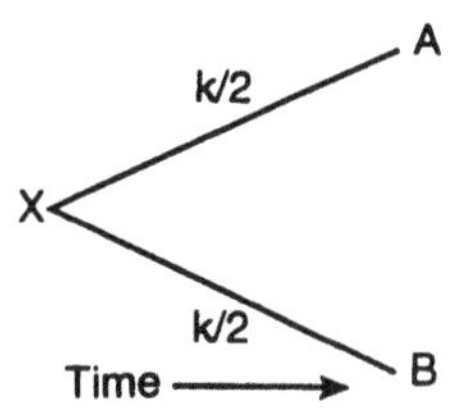


Let us apply this method to cytochrome c, a protein of the cell respiratory chain. Its sequence has been determined in many organisms. It is 104 amino acids long in all animals. When we compare cytochrome c from the human and rhesus monkey, we find just one difference, in position 66; the 103 other amino acids are identical. We find 12 differences when we compare cytochrome c from the horse and human but only 11 differences between the horse and monkey. These results are represented in the "non-rooted" tree on the left (p. 128). If we assume that the substitution rate is more or less the same in all branches, we can conclude that the lineages of man and of rhesus monkey diverged later than those of man and horse, and that the difference of one amino acid in the cytochrome c sequences of man and monkey appeared after these two species had separated. The middle rooted tree is the one that assumes the greatest equality in substitution rates. The tree below right has to be rejected.

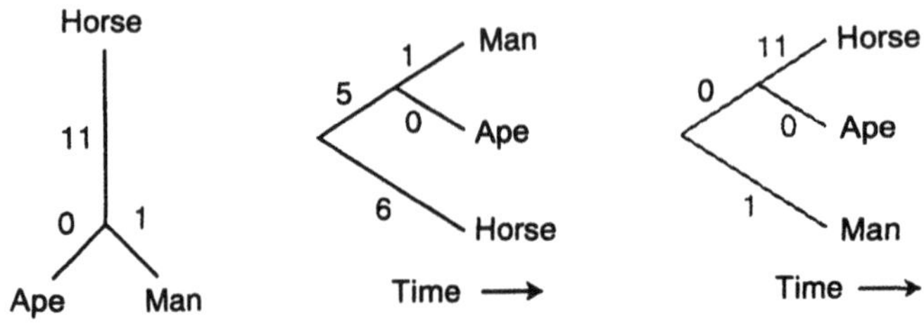


The methods for estimating genetic change throughout evolution include the hybridisation of DNA molecules of different origin, protein sequencing, immunological analysis and electrophoresis. Comparisons of the protein sequences of a variety of organisms have confirmed that, in many cases, the evolution rate of protein sequences is remarkably constant. Thus, the sequence of haemoglobin α has evolved at the same rate in all vertebrate lineages, whether these lineages have undergone pronounced and rapid morphological and ecological evolution (hominids or ungulates compared to other mammals or, at a higher taxonomic level, amniotes compared to other vertebrates) or whether they have not evolved, or hardly evolved, morphologically for a long time (selachians, salamanders or, amongst mammals, opossums). This apparently constant rate of evolution of a given protein or DNA sequence has led to the molecular clock hypothesis (E. Zuckerkandl and L. Pauling, 1965). The existence of such a clock has been a strong argument in support of the neutral theory, even if its regularity among lineages, as its causes, are still subjects of controversy.

**The technique of DNA/DNA hybridisation**

This technique evaluates the overall similarity in the DNA of different organisms. The starting material is the total DNA of a given species. The repetitive DNA sequences present in the genome of most higher organisms are eliminated. The sample of the unique - most informative and most specific - DNA sequences thus obtained is then labelled with radioactivity and "denatured" (the two strands of the double helix are separated, for instance by increasing the temperature). The DNA preparation is then fractionated and left to react with denatured DNA from another species. If the two genomes placed together have similar sequences, the complementary strands of different origin will pair up to form hybrids. The extent of the reaction gives the proportion of similar DNA sequences.

Not all the nucleotides of the sequences that form hybrids need to be complementary. The level of non-complementary nucleotides in the DNA hybrid is given by the rate at which the paired strands dissociate when the temperature is increased. The thermal stability is the temperature at which 50 percent of the DNA hybrid is dissociated. The difference in thermal stability of hybrids from different pairs of species can be used as a phylogenetic character.

The principle of the molecular clock is the following. If molecular evolution rate is really constant over time, by dating stages of evolution by other methods (e.g. palaeontology) we can calibrate the chronology of a protein's or sequence's phylogenesis. The neutral theory does not, however, claim to offer a perfect timekeeper but rather a kind of stochastic clock like the one that measures a decrease in radioactivity in physics. It is the probability of change that is constant. Despite some variability the molecular clock can be considered to be fairly accurate over a long time period. Under certain circumstances, however, it becomes most imprecise. A sequence can be stable in one species but evolve very rapidly in another. Faced with this kind of situation, we assume that either the mutation rate itself differs among species or that selection acts differently according to species. For example, some alleles could be neutral in one species and selected in another.

Evolution rate may well be more or less constant among species for a given protein but it varies considerably among proteins. (It is 1.2 substitutions per amino acid and per one thousand million years for the $\alpha$ chain of haemoglobin, 0.3 for cytochrome c, 0.01 for histone H4, but 8.3 for fibrinopeptide).[33] This does not mean that fibrinopeptide undergoes more mutations than histone H4 but that the proportion of neutral mutations is very different. For fibrinopeptide, the lethal mutation rate $\mu$ is lower than the neutral mutation rate $\nu$. For histone, on the contrary, $\mu$ is greater than $\nu$. Histone H4 has virtually the same sequence in all eukaryotes, from men to peas. The gene does mutate but any mutation affecting its primary structure is strongly counterselected because of the strong functional constraints on the protein. (Histones take part in the packaging of DNA in chromosomes.) On the contrary, the only function of fibrinopeptide is to prevent, by its presence at the N-terminal end of plasma fibrinogen, the spontaneous polymerisation of fibrin into a mesh (coagulation). The precise sequence of its amino acids is therefore unimportant and many of the mutations are neutral.

Similarly, some DNA sequences evolve faster than others. For example, introns (non-coding regions of genes that interrupt the coding sequence) evolve faster than exons (regions coding for genes between which introns squeeze themselves) because a change in sequence is less likely to affect the expression or structure of the gene product of introns than exons. Moreover, a comparison of coding sequences in species that are close shows that the rate of substitution of bases is higher in the third position than in first two positions of codons. (Because there are synonymous codons that only differ in the third position, this base can often be substituted without altering the meaning of the genetic message.) In general, the non-coding regions of a genome evolve faster than non-coding regions.

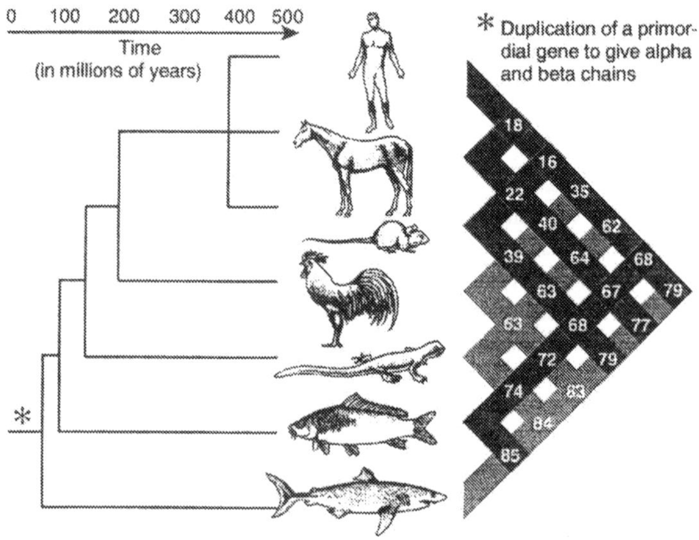


*Figure 3.13.* Phylogenetic tree showing the evolutionary relationships among seven vertebrates, that is the order in which their lineages diverged from each other during geologic time and the dates when this occurred. To create this tree, we studied a key protein, α haemoglobin, which differs among the animals of the sample. The table on the right gives, at each cross-section, the number of amino acids that differ in the sequences of the α chains of two animals. The α chain is close to the β chain, the other subunit of the haemoglobin molecule (which is formed by the assembly of two α chains and two β chains). The two chains were formed by duplication of an ancestral gene 450 million years ago. The table shows that the evolution rate of one protein in several very different organisms is fairly uniform (as predicted by the neutral theory). The number of differences depends only on time elapsed since the two lineages separated. For example, the number of different amino acids is close to 20 in two of the three mammals and close to 70 between the carp and each of these two species (black boxes).

Each gene or protein thus ticks at a different rate (at the mutation rate, v, of the gene's neutral alleles) but all genes provide a similar date for the same evolutionary event.[34] This statement was first thought to contradict Darwinian orthodoxy. If we accept that most evolutionary changes are adaptive, the molecules and sequences crucial to organism adaptation should be those that evolve fastest. In fact, in neutral theory, the role of natural selection in the evolution of organisms is not disputed just as polymorphism selection is not disputed in sickle-cell anaemia. The neutralist hypothesis just implies that most of the polymorphism highlighted by electrophoresis or the study of DNA is neutral and that cases of selected polymorphisms are few. Similarly, within a lineage of proteins, some substitutions undoubtedly do have adaptive significance. For example, we could probably not exchange the $\alpha$ haemoglobin chain of carp and man with impunity. Several of the 68 amino acids that differ between them might play a role in aspects of oxygen transport that are species specific. However, these adaptive substitutions are drowned by the volume of neutral substitutions with no adaptive value.

The debate between "selectionists" and "neutralists" which emerged in the early 1960s was very heated in the 1970s and 1980s and spurred many studies in population genetics. It is, however, impossible to prove that a polymorphism is neutral (and difficult to prove that it is not). How can we be sure that we have eliminated all selective factors? At all events, neutralists' calculations showed that their hypotheses could account for most observations and could also make predictions regarding molecular evolution and the structure of variability which were apparently correctly. In the 1950s-1960s, many evolutionists had even refused to consider that there might be such a thing as a neutral mutation. Ernst Mayr thought it highly unlikely that a gene could remain selectively neutral for long.[35]

Nowadays, many partisans of neo-Darwinism are more willing to give due consideration to neutralism. In 1984 J.H. Gillespie wrote:

> *...acceptance of the neutral theory seems to have increased considerably. This theory is now invoked as routinely as selection was a few years back.*[36]

Like R.C. Lewontin, who remains highly critical of the neutral theory, most evolutionists accept that much protein and gene polymorphism could be neutral even if they still disagree on the importance of chance and selection.

## NOTES

1. The terms "gene", "allele" and "locus" have been used in the previous chapters and, from now on, will be used regularly. Their definitions, and those of other terms often used in genetics (*indicated by asterisks*), are given in the glossary.
2. Mutation can refer to both process and outcome.
3. We shall come back to transposable elements later. These DNA sequences can move within the genome. Most contain genes coding for their own transposition. Mutations due to transposable elements are an important source of genetic variability.
4. Monod, J., 1970, *Le Hasard et la Nécessité*, Seuil, Paris, p. 120.
5. Gros, F., 1986, *Les Secrets du Gène*, Odile Jacob/Seuil, Paris.
6. Watson, J.D. and Tooze, J., 1981, *The DNA Story. A documentary history of gene cloning*, W.H. Freeman & Co, San Francisco, p. vii.

7. Darwin, C., 1859, *On the Origin of Species by Means of Natural Selection*, Murray, London.
8. Clark, A.G., 1989, *Principles of Population Genetics*, $2^{nd}$ ed, Sinuaer Associates, Sunderland, Mass., p. 306.
9. The founder effect explains the high rate of genetic diseases in groups arising from small numbers of migrants (for example, Tay-Sachs disease amongst the Ashkenazim of Quebec).
10. Limits have to be fixed for the inbreeding coefficient to work and so that it does not always equal 1. For example, we reduce it to 0 when there is a mutation.
11. Cann, R.L., Stoneking M. and Wilson, A.C., 1987, Mitochondrial DNA and human evolution, *Nature* 325: 31-36.
12. See Lewin, R., 1991, l'Anthropologie moléculaire, *La Recherche*, 22 (N° 236): 1249.
13. See Klein, J., Takahata, N. and Ayala, F., 1994, Reconnaissance du soi et origine de l'homme, *Pour la Science*, 196, p. 60.
14. See Bourguignon, A., *L'homme Imprévu* (1989) et *L'homme Fou. Histoire Naturelle de l'Homme* (1994) (2 vols), Questions, Presses Universitaires de France, Paris.
15. Fisher, R., 1950, (originally published in 1926), *Contributions to Mathematical Statistics*, Wiley, New York.
16. Before turning their punctuated equilibria into unexplained phenomena, N. Eldredge and S.J. Gould had offered this explanation (see Chapter 7).
17. Haldane, J.B.S., 1938, *Forty Years of Genetics*. In: *Background to Modern Science*. Ed. J. Needham and W. Pagel. Cambridge University Press, pp. 225-243.
18. It is rather amusing that Fisher, a conservative, and Haldane, a member of the central committee of the British communist party, hardly disagreed on anything in science.
19. Ford, E.B., 1975, *Ecological Genetics*, Chapman & Hall, London; Kettlewell, H.B.D., 1973, *The Evolution of Melanism*, Clarendon Press, Oxford; Bishop, J.A., and Cook, L. M., 1987, Industrial melanism and the urban environment, *Adv. Ecol. Res.* 11: 373-404.
20. Clarke, C.A., Mani, G.S., and Wynne, G., 1985, Evolution in reverse: clean air and the peppered moth, *Biol. J. Linn. Soc.* 26: 189-199; Brakefield, P. M., 1987, Industrial melanism: Do we have the answers? *Trends Evol. Ecol.* 2: 117-122.
21. Bishop, J.A., 1972, An experimental study of the cline of industrial melanism *in Biston betularia (L.) (Lepidoptera)* between urban Liverpool and rural North Wales, *J. Anim. Ecol.* 41: 209-243; Mani, G.S., 1980, A theoretical study of morph ratio clines with special reference to melanism in moths, *Proc. Royal Soc. London*, B210: 299-316; Brakefield, P. M., *op. cit.*
22. Creed, E.R., Lees, D.R., and Bulmer, M.G.(1980) Pre-adult viability differences of melanic *Biston betularia (L.) (Lepidoptera)*, *Biol. J. Linn. Soc.* 13: 251-262.
23. We shall address this hypothesis at greater length in "Struggle for life and a royal race" (Chapter 7, Section 5).
24. Let's not forget that some species are no doubt more diverse than others. The species is thus not a suitable unit to measure biodiversity.
25. Darwin, C., *ibid.*
26. Muller, H.J. et al., 1939, The Geneticists' Manifesto, *Nature* 30: 371-373.
27. It may seem surprising that the hybrid benefits when advantageous genes combine yet does not suffer when unfavourable genes combine. Voltaire retorted to the marchioness who suggested they should have a child combining her beauty and his great intelligence that he feared the reverse might happen. In general, favourable genes are dominant either because unfavourable genes are often inactive - and inactivity tends to lead to recessiveness - or because a "good" gene that has become dominant by mutation will be selected.
28. It is a question of polymorphism, and not polytypism, when individuals from different habitats mate. We are dealing with a single population (as in plants belonging to a

particular habitat, but subject to genetic exchanges between habitats through pollen and seed dispersal.)

29. Robert, J.M., 1994, Le fardeau génétique, APBG (*special issue*), p. 45.
30. In the case of overdominance, it is not possible to maintain 100 percent of heterozygotes by sexual reproduction. Because of random allele segregation, homozygotes, which are not advantaged, reappear at each generation.
31. Kimura, M., 1983, *The Neutral Theory of Molecular Evolution,* Cambridge University Press, Cambridge.
32. The comparative anatomy of a yeast, pine and human is not easy to establish nor really relevant unlike the comparison of their macromolecules.
33. Histone H4: A protein that is part of the structure of the eukaryote chromosome. Fibrinopeptide: A polypeptide chain that is a product of blood coagulation. It is obtained, together with fibrin, by the enzymatic cleavage of fibrinogen by thrombin.
34. Ayala, F.J., 1982, *Biologie Moléculaire et Evolution.* Series: Biologie évolutive, Masson, Paris, p. 79.
35. Mayr, E., 1963, *Animal Species and Evolution,* Belknap Press, Harvard University Press, Mass.
36. Gillespie, J.H., 1984, The status of the neutral theory, *Science* 224: 732-733.

# Chapter 4

# THE KEY TO GENES

Nature studies in the 17th and 18th centuries had underscored the conflict between co-existing species. In the 19th century, Darwinism had highlighted the conflict among individuals of a single species. At the turn of the 20th century, what was left of the idea dear to Ancient philosophers that independence and cooperation lie at the source of natural harmony? It had taken refuge in the individual. An organism was defined as an ensemble of organs which were adapted to a particular lifestyle and which, by cooperating, ensured the individual's survival and the highest reproduction rate. Although modern biologists had accepted the idea of selection, this idea was still subject to the view that the individual is a harmonious unity. Many biologists and others continued to think of the living world in terms of two fundamentals: the hierarchy of biological systems (part of an organism could not evolve at the expense of another) and the shaping of an ever better world (by improved adaptation). This fragile compromise - and it was a compromise - centred on the individual exploded in the second half of the 20th century.

The discovery of DNA by J. Watson and F. Crick in 1953 added a new molecular dimension to our understanding of life. A special kind of information transfer occurs at the DNA level; it lies at the heart of vital processes and explains morphogenesis and reproduction. It prompted the questions: On what entities and at what integration levels does natural selection act? Must we stick to Darwin's struggle for life - the struggle between individuals - and preserve individual harmony? However, if we read Darwin's *Origin of Species* carefully, we find that his thoughts on this subject were not as clear-cut as we may have believed at first. His theoretical framework can embrace the idea of selection acting at different levels and even within each organism. Neo-Darwinians thus undertook to expand on Darwin's view by taking into account supporting knowledge from genetics and on selection mechanisms to explain the living world. For a neo-Darwinian, no integration level can escape selection.

This insight has radically changed the way we view life. Not only is the definition of life but also of the individual beyond the investigative powers of human intelligence (see Introduction)! However, the step taken backstage by the individual is offset (and also provoked) by the arrival onstage of a new concept, information. The concept of information helps us to clarify the links between genetics and evolution, and to elaborate a view of life integrating all levels, from molecule to group.

This chapter will explain this new outlook. We shall first introduce the main themes - individual, information and selection levels using as an illustration the presence of females within a hermaphrodite population, then address each of these themes from a theoretical viewpoint.

## 1. MALE OPPOSITION

Selection occurs at the population level because there is just not enough room for an infinite number of individuals. If selection did not occur, even individuals which reproduce less well than others would be maintained. Something similar happens at the individual level. An organism, for example a plant, acquires a restricted amount of resources within a given time span. These are allocated to a variety of functions. That which is allocated to one function, however, cannot be allocated to others; in other words, there is a trade-off. At any moment, the plant can use its resources for either growth (to have more resources later on), survival, or production of sexual (pollen or seeds) or asexual propagules. Agricultural experts are well aware of this trade-off among functions and try to take advantage of it.

The allocation of resources for reproduction poses interesting problems but, first, we must distinguish between male and female functions. These can co-exist in a single individual.

**The gender of flowers and plants**

Botanical terminology relating to sex, introduced in part by Darwin himself, is not very clear as it does not distinguish levels (flower, individual, population and species). We shall therefore define each term we use unambiguously. The flower can be hermaphrodite (i.e., carry male organs (stamens) and female organs (pistil, ovary)), female (or sterile male, we shall see why below), or male (sometimes, but rarely, denoted sterile female). The basic floral type is the hermaphrodite. Individual plants are either hermaphrodites (all flowers are hermaphrodites), monoecious (both male and female flowers), andromonoecious (male and hermaphrodite flowers), gynomonoecious (female and hermaphrodite flowers) or trimonoecious (all three types of flowers). At the species level, we speak of either hermaphrodism (all individuals are hermaphrodites), monoecia (all monoic), dioecia (some male, some female individuals), gynodioecia (females and hermaphrodites), androdioecia (males and hermaphrodites) and trioecia (all three types of individuals).

In their 1976 attempt to model the male and female functions, E.L. Charnov, J. Maynard Smith and J.J. Bull[1] concluded that the resources allocated to both male and female must be equivalent whether the species is hermaphrodite, monoecious (in both these cases, male and female functions are united in a single individual) or dioecious. This conclusion seems counter-intuitive because, by definition, the male function produces smaller gametes than the female does. We would expect investment into the female function to be of greater value for parental fitness. Even if we accept that a dioecious species such as ours squanders its reproduction potential by producing about as many males as females, we accept less readily that the pollen seeds (carriers of male gametes) of a hermaphrodite or monoecious plant are as costly as its seed-laden pods.

Equal allocation of resources to males and females only holds for allogamous plants (which require another individual to reproduce; a hermaphrodite or monoecious plant is allogamous if it cannot self-fertilise). Autogamy (self-fertilisation) restricts "squandering". Selection gives the advantage to the autogamous individuals that invest least in their male function; the plants do not compete to fertilise a given flower. On the other hand, allogamous individuals do compete in order to reproduce, and there is no restriction on squandering. Thus, when we select the female function - the seed - in cultivated allogamous plants (if we are not going for growth or reserves), we do not reduce investment in the male function. The African millet producer who selects the plants with the largest number of seeds privileges the genes of the mother-plants as much as of the father-plants which provided the pollen; selection is as strong on male and female functions.

In gynodioecious species, which are fairly widespread and comprise both female (or sterile male [2]) and hermaphrodite (or fertile male) plants, the reproductive performance of females far outweighs that of hermaphrodites. For example, in plantain, females produce about 70 percent more seeds than hermaphrodites do; in thyme, they produce twice as many. Thyme, which we have already used as an example in Chapter 3, is an excellent illustration of gynodioecia for other reasons also:

1. Females often appear in hermaphrodite species because the male function, like all other biological functions, can be lost. In most hermaphrodite species, male sterility is just a very rare genetic disease. It is not due to genuine polymorphism but, at most, to what we have called cryptopolymorphism. In thyme, however, the proportion of females in wild populations is high, always above 10 percent, and sometimes even above 90 percent. For instance, near Montpellier, the mean proportion of females is 60 percent and that of hermaphrodites 40 percent. No male has ever been found. It is thus highly likely that thyme genes will be carried by a female plant as they have been selected in a context of gynodioecia.

2. When female (sterile male) plants make up a large part of the population as in beans, a plant's phenotype may be difficult to establish. Bean populations often contain plants of intermediate phenotype. On the other hand, thyme plants can nearly always be classified unambiguously as either female or hermaphrodite. Although there are several ways of being female (the stamens can be more or less aborted) and only one way of being hermaphrodite, the two forms are easily distinguished (see Fig. 4.1).

*Figure 4.1.* Thyme flowers from a female (left) and hermaphrodite (right) plant. The female flowers do not have stamens. In the hermaphrodite, the stamens develop before the pistil (not shown).

3. In some species displaying frequent, stable male sterility, female production capacity in hermaphrodites is so reduced that we can hardly speak of gynodioecia. The species are potentially but not functionally gynodioecious. The fig tree is a good example. Its hermaphrodite form does

not produce seed but "raises" pollinators in its female flowers. Asparagus may be yet another example. Hermaphrodism is so limited that the plants are said to be males even though they can produce seed under certain conditions; female asparagus plants are strictly sterile-male. The hermaphrodites of both the fig tree and asparagus may not have lost their potential to produce seed but they have lost the genetic or ecologic equipment needed to express this potential. Both species illustrate the passage from a hermaphrodite to a dioecious state. The case of thyme is not as simple. According to genetic and ecological studies, the wide range of sex ratios in populations of wild thyme corresponds to a great diversity in the genetic information determining gender. Interbreeding of thyme plants of different origin has led to disjunctions in sex characters. Their complexity cannot be explained without reference to nuclear and cytoplasmic genetic determinants (see below).

It is not clear how male sterility genes can propagate in a hermaphrodite population to the point of overrunning it, nor how hermaphrodites turned female by losing their ability to produce male gametes maintain themselves. For a start, it is hard to consider genodioecia as an evolutionary end-point. An evolutionist would consider it a transition step or an intermediate stage between the hermaphrodite and dioecious states. However, as Darwin noted, the gynodioecia of thyme displays all the properties of a stable state. Is this a kind of stable disequilibrium?

In *The Different Forms of Flowers in Plants of the Same Species* (1877), Darwin used thyme as an example to ask how plants that only reproduce through the female route could be selected instead of hermaphrodites which reproduce through both male and female routes. In 1941, D. Lewis' work provided part of the answer to the question by showing that a male sterility gene does not undergo the same evolutionary constraints if it is nuclear (present in the cell nucleus and able to be passed on by either gamete) or cytoplasmic (included in the genome of organelles - mitochondria or chloroplasts - and passed on by the maternal route). Selection acts quite differently on nuclear and cytoplasmic genes because of the distinct ways in which they are passed on.

Let us consider first the more "conventional" nuclear genes inherited both paternally and maternally. These genes are reproduced far better by hermaphrodites than by females. Each time that a female produces a seed, it reproduces half of its own genome and half of the genome of the hermaphrodite that fertilised it. Females thus contribute toward the reproduction of the nuclear genes of hermaphrodites. The reverse does not apply because hermaphrodite seeds only contain hermaphrodite genes. The genes of females thus reproduce far less well than those of hermaphrodites (see Fig. 4.2). Ultimately, these genes - and females - should disappear (if gender is coded at the nuclear level).

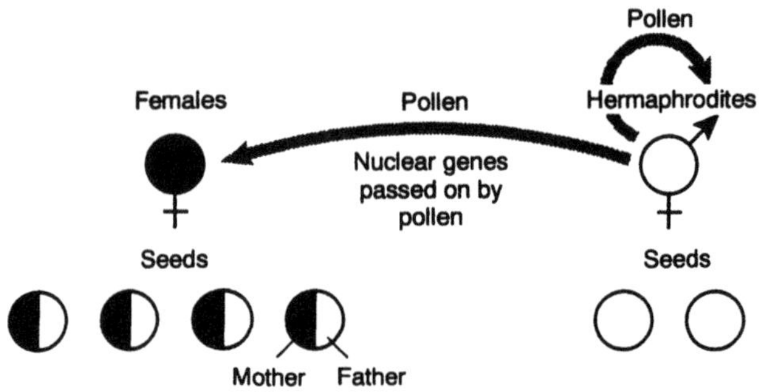


*Figure 4.2.* Nuclear genes (inherited from both parents) are passed on less well by females than by hermaphrodites. Let us imagine a population of an equivalent number of females and hermaphrodites where females produce twice as many seeds as hermaphrodites. Each seed produced by a female (after fertilisation by the pollen of a hermaphrodite) reproduces the nuclear genes of the female and of the hermaphrodite. Each seed produced by a hermaphrodite only reproduces the genes of the hermaphrodite. As a result, in the six seeds that are produced, the genes from the hermaphrodite are multiplied eight times (twice maternally and six times paternally) whereas the genes from the female are multiplied only four times. Thus, the genes of females are reproduced half as often.

Now let us consider the cytoplasmic genes in cytoplasmic organelles which are all - or almost all - inherited maternally. This is a quite different case. Because these genes are not passed on by the male, it is immaterial whether their carrier produces pollen or not, as this will not change their transmission mode. Ultimately therefore, the cytoplasmic genes of females should overrun the population because they produce more seeds (see Fig. 4.3).

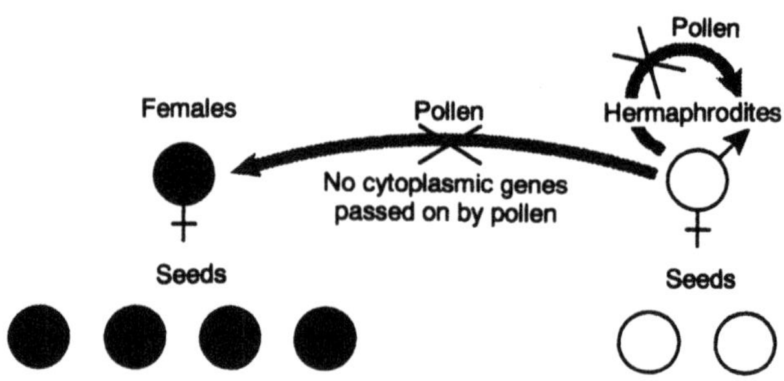


*Figure 4.3.* Cytoplasmic genes (inherited maternally only) are passed on better by females than by hermaphrodites. The population is the same as in Fig. 4.2; there are as many females as hermaphrodites and the females produce twice as many seeds as the hermaphrodites. In the six seeds that are produced, four contain the genome of a female whereas only two contain the genome of a hermaphrodite. The genomes of females thus reproduce twice as much as those of hermaphrodites.

We encounter both hermaphrodite and female predominance in thyme. In some populations, at certain times, the difference between hermaphrodites and females is chiefly due to a difference in cytoplasmic genes. Cytoplasmic genes in females (which determine the female form) increase in frequency, leading to an increase in the proportion of females; this increase can sometimes be very high. At other times, hermaphrodites and females differ primarily in their nuclear genes. Nuclear genes determining the female form are not advantaged, and the proportion of females decreases. Nuclear and cytoplasmic genes are subject to opposite selective forces here, and we speak of "conflict" (we shall come back to this later). Nuclear genes are selected to produce hermaphrodites. Investment in the male and female is equal because nuclear genes are passed on equally by both routes. The cytoplasmic genes are selected to produce females, and all is invested in this route which is the only one to reproduce them. According to molecular and theoretical analyses, the "feud" between nuclear and cytoplasmic genes is endless in thyme and in many other plants. Cytoplasmic genes cut off male function, nuclear genes restore it.

The genodioecia of thyme thus puts us in an embarrassing but instructive quandary. The question "Does selection advantage females or hermaphrodites?" has no answer because it is incorrectly phrased. For an answer, we must change the question to either:

"Which nuclear genes does selection advantage, those that determine the female or the hermaphrodite form?" The answer is that selection advantages the nuclear genes that determine the hermaphrodite form;

or "Which cytoplasmic genes does selection advantage, those that determine the female or the hermaphrodite form?" The answer is that selection advantages the cytoplasmic genes that determine the female form.

But, if this is the case, selection does not act upon the individual but upon the gene! As long as the two entities are confused, the question goes unnoticed. We have stumbled upon it by accident just because, in the example of the gender of thyme, the reproduction of individuals and that of genes depends on the type of gene (nuclear or cytoplasmic) considered. Genes, not individuals, are reproduced. Does this mean that the history of life is just the history of genetic information passed on from one generation to the next by individuals with an ephemeral existence? We shall discuss this now and return to gene "conflict" in an organism later.

## 2. INDIVIDUAL, WHERE ARE YOU?

Back in 1945, G. Östergren, a Swedish biologist, postulated that:

*the units of selection are never biological individuals but their genes and their chromosomes.*[3]

This was a daring statement. Is not considering genes or chromosomes as possible selection levels endowing them with a status normally reserved for the individual, organism or living creature?

To discover the relevance of Östergren's postulate, we shall first provide a few definitions starting with that of an individual. According to its Latin etymology, an individual designates a reality that is:

1. indivisible, from a given standpoint, thus guaranteeing its unity,
2. singular, different from any other, thus guaranteeing its uniqueness.

Indivisibility and uniqueness are the two fundamental properties of the individual. The commonly accepted view of individuality is modelled, to a large extent, on life and its internal finality; in biology an individual is an entity that displays genetic and organic unity (or indivisibility) and that is singular because it is inscribed in history. For example, a zygote is an individual because it is genetically unique, has organic and genetic unity (despite being a simple cell, it has regulatory mechanisms), and because it is a singular (and primitive) step in ontogenesis.[4]

Uniqueness, unlike indivisibility, poses problems. At the root of the astonishing and provocative statement that the zygote is an individual lies an epistemological difficulty of ontological origin. If the aim of science is to generalise and reduce to the identical in order to understand and explain, must we conclude that the individual is the limit of what can be scientifically known? Is a science of the individual not a contradiction in terms? *Individuum ineffabile est*; there is nothing to say of the individual, declared Thomas Aquinas. Plato's Ideas and their "participation" in the material world thus resurface in the question: Are there non-individuated realities that could belong to several individuals?

It is hard to find answers to these crucial questions because we are still influenced by the models inherited from Aristotle and Descartes. These models are mainly of a physiological and "technological" sort, and have an instrumental, utilitarian outlook. The first question asked is "What is it for?" There are, however, other ways of looking at things. For example, to the philosopher Leibniz, the cell, which had just been discovered by Leeuwenhoek, was not a tool meant for a particular organ but an individual, an organism with functions, a "monad". In *Monadology*, Leibniz defined monad as "*a simple substance, without parts, that enters into the composition of compounds*"; "*these monads are the true Atoms of Nature, in a word the elements of things*". All monads differ and change continuously from within.

According to monadism, the universe is made up of these well-defined units which belong to a spiritual order unlike atoms which belong to the world of mechanics. The monad-cell is not separate from the Universe, nor is it a specific element of a fundamental and causal whole. It represents the whole (it is sometimes called a profile) and shares the fate of the Universe. We shall discuss the limitations of this viewpoint incarnated by the eminent Dr. Pangloss below. However, if the cell is an individual, linked to other individuals (cells) by a higher-order relationship, then the scale used in a physiological approach must be reduced to that of a cell. It ensues that the structure of the individual (organism) becomes subordinate to the functions of the component, that is of the individual (cell). Made up of cells, the organism is "made for" cells, for components that are themselves wholes, albeit less complex ones. But if an understanding of the whole depends on an understanding the parts, are we not being reductionist?

Leibniz's view - the transition from the universe to the monad - prefigures in part how biological thinking was to evolve in the 20th century. However, even if early 20th century biologists could cut living things up into smaller pieces than either Leibniz or his contemporaries could, they did not discard Aristotelian and Cartesian models completely. The criticisms waged against Östergren's hypothesis are proof of this. Nevertheless, two major conceptual upheavals of the second half of the 20th century will consolidate the breach made by Östergren, namely, asking "what is life?" in physicochemical terms and the advent of the notion of information.

## 3. WHAT IS LIFE? ERWIN SCHRÖDINGER'S "ESSENTIALIST" ANSWER

There is no doubt that physical and chemical sciences exert an influence on natural sciences. The scientific revolution of the 17th century resulted in the prevalence of a determinist approach to reality and to the study of living things. Since then, biochemistry and molecular biology have reinforced the influence of the "hard" sciences. At the same time, however, emergent scientific fields have challenged the dominance of determinism. Examples of these new fields are thermodynamics, a science of the macroscopic world which states that nature's laws are probabilistic, and the science of chaos which focuses on all that is irregular, random, or turbulent. Although biology won its independence in the 19th century in a fight against a certain kind of physics, it has also benefited from this discipline's constant self-appraisal. In the first half of the 20th century, the sciences of matter underwent deep changes which were to inspire a great theoretical physicist's thoughts on the material basis of life.

The Austrian physicist Erwin Schrödinger is the author of the famous wave equation of quantum mechanics. This equation was named after him and, in 1933, earned him the Nobel prize which he shared with a fellow scientist, Paul Dirac. Schrödinger lived at a turning point in the history of sciences which was characterised by changing relations between physics and biology. Schrödinger focused on these relations and, to his contemporaries who were astonished that a physicist dared write about a scientific field in which he was not qualified, he replied that the nature of his enquiry was universal and rooted in the philosophy of his culture. He was, he said, examining life and the living from his own philosophical standpoint using his own scientific method, constructivism. Concepts and scientific facts were to be construed as constructs of the human intelligence and not as data established by nature.

Before tackling the question "*What is life?*" (the title of his essay[5]), Schrödinger asked himself:

> *how can events in space and time which take place within the spatial boundary of a living organism be accounted for by physics and chemistry?* [6]

It is a legitimate question because today, and even more so in Schrödinger's day, physics and chemistry have been unable to come to grips with most vital phenomena, in particular with the transition from inert to living matter, from the inanimate to the animate. However, for Schrödinger, this "*obvious inability of present-day physics and chemistry is no reason at all for doubting*" that these sciences will not be able to account for such phenomena in the future.[7] Living beings do not "*elud(e) the laws of physics as established up to date*" even if these laws cannot explain the specificity of living matter which must obey "*other laws of physics hitherto unknown, which, however, once they have been revealed, will form just as integral a part of this science as the former*".[8] Inadequate analytical methods and especially biological complexity are hurdles that have yet to be overcome.

We may find Schrödinger's hegemonic view of physics rather extreme. There is no justification to keep calling the search for knowledge that has been extended to include living beings "physics"; it is just "science". In fact, Schrödinger did not suggest that natural selection, which the biological community had accepted long ago, should be part of his physics.

**Physicists and the issue of life**

Schrödinger took up a tough challenge. Several of his contemporaries were convinced that physics and chemistry could not explain life. Their outlook had, in fact, helped biology gain the independence that it so sorely needed and that vitalism had defended in its own way a century earlier. According to J.B.S. Haldane, the more discoveries we make about physiological activity and heredity, the more difficult it becomes to conjure up a description or an explanation in physical or chemical terms that can embrace all the facts of persisting coordination.[9]

The physicist Niels Bohr was less assertive than Haldane. He recognised the specificity of vital processes and wondered whether a physicochemical explanation was possible. At the same time, he underscored analogies with the new physics. Just as life is not explained by biology, quanta are not explained by quantum mechanics. They are, on the contrary, the basis of quantum mechanics:

*The recognition of the essential importance of fundamentally atomistic features in the functions of living organisms is by no means sufficient, however, for a comprehensive explanation of biological phenomena. The question at issue, therefore, is whether some fundamental traits are still missing in the analysis of natural phenomena, before we can reach an understanding of life on the basis of physical experience. [...] On this view, the existence of life must be considered as an elementary fact that cannot be explained, but must be taken as a starting point in biology, in a similar way as the quantum of action, which appears as an irrational element from the point of view of classical mechanical physics, taken together with the existence of the elementary particles, forms the foundation of atomic physics.*[10]

The first question Schrödinger asked in his essay was why are atoms so small. This question, he said, is meaningless or, rather, has meaning only in relation to man's measure, thus echoing Protagoras' words, "*man is the measure of all things*".[11] By commenting on atom size, Schrödinger wanted to point out that the strange properties of the subatomic world revealed by the new physics were no reason for not using physics in biology. The physicochemical laws to be applied to living things were of a solely statistical nature; atoms move incessantly and unpredictably. However, his reasoning had shortcomings. The physicochemical laws must apply not only to a living being as a whole, that is to an entity of an astronomical number of atoms, but also to groups of atoms that are too small for the law of large numbers to apply. These small groups of atoms are important because they determine the "order" in which organisms appear, grow and die. They are the chromosomes.

Schrödinger's thinking accompanied a scientific revolution - the advent of molecular genetics - that marked the end of the first half of the 20th century. Molecular genetics took a strictly mechanistic and reductionist view of living organisms and, instead of investigating functions or organic parts as a whole, studied the structures of their universal components. A typical approach was taken by a pioneer of molecular biology, Max Delbrück, who chose to work on the most simple material, the bacteriophage (a virus that infects bacteria). The mechanisms that he discovered in this rudimentary organism as regards both structure and function were so fundamental that the results could be extended to all living things. Jacques Monod, who was very close to the "phage group" headed by Delbrück, later wrote:

*What do, for example, a blue algae, an infusoria, an octopus and man have in common? The discovery of the cell and cell theory highlighted a new unity within this diversity. However, not until the developments in biochemistry of the second quarter of the 20th century was the deep, strict unity of all the living world revealed at the microscopic level. We now know that the structure and mode of operation of the chemical machinery are intrinsically the same from the bacterium to man.*[12]

The unity of the living world at the molecular level must be compatible with substantial diversity because, as Schrödinger writes, at this level are coded:

*not only the structure and functioning of that organism in the adult, or in any other particular stage, but the whole of its ontogenetic development from the fertilized egg cell to the stage of maturity, when the organism begins to reproduce itself.*[13]

This led him to suggest that an organism has a "pattern". The four-dimensional pattern (the three dimensions of space plus time) would be repeated from one individual and one generation to the next. The implication was that the concept of information embracing a message and a possible feedback could be applied to living things. Structural rigidity and conformity to a specific type depend upon the execution of a strict, prescribed programme, whereas flexibility in behaviour is guaranteed by self-regulatory mechanisms and feedback loops. As François Jacob says:

*All organisations involve feedback loops in which each element is kept informed of the effects of its function which it adjusts in the interest of the whole.*[14]

As early as in 1944, Schrödinger used this "second-hand" biological knowledge to predict some of the main physical properties of a chassis for heredity. They were not confirmed until 1953, the year that J. Watson and F. Crick described the double helix structure of DNA. Schrödinger's predictions were a brilliant demonstration of the effectiveness of constructivism. Moreover, the demonstration gave Einstein an *a posteriori* reason for satisfaction since, in disagreement with Heisenberg, he claimed that only theory determines what can be observed. Schrödinger concluded that, given the prerequisites of a chassis for heredity, the only possible arrangement of atoms was the one found in a molecule with a regular, aperiodical crystal-type structure (with no repetitions). The gene could therefore not be a combination of atoms that owes its properties to the laws of statistical physics.

> *... a gene contains certainly not more than about a million or a few million atoms. That number is much too small [...] to entail an orderly and lawful behaviour according to statistical physics - and that means according to physics. It is too small, even if all these atoms played the same role, as they do in a gas or a drop of liquid. And the gene is most certainly not just a homogeneous drop of liquid. It is probably a large protein molecule, in which every atom, every radical, every heterocyclic ring plays an individual role, more or less different from that played by any of the other similar atoms, radicals, or rings.*[15]

Schrödinger advanced several arguments to support his thesis that heredity is determined by a very small number of atoms. First, the message is stable because the changes on an atomic scale revealed by quantum mechanics are discontinuous. The "aperiodical crystal" he imagined would lose its stability only as a result of a strong perturbation, a quantum jump to a different energy level, in other words a mutation. Second, an arrangement of several motifs at the molecular level could account for the variety demanded by life's diversity. Just a few motifs would be enough. Does not Morse or binary code in computers encode any text with just two symbols? The order in vital processes would thus derive from the order in chromosomes to which would be added the statistical, physicochemical regularity of molecule behaviour in a cellular environment.

But did Schrödinger locate life for us? Probably not. We now accept that life can neither be reduced to one or more structures nor observed in the laboratory with, for example, the tools of molecular biology. The originality and ambiguity of biological events lie in their halfway house position between the manifestations of inert matter (determined matter, Schrödinger would have said) and thought. At the end of his essay, Schrödinger transformed his *What is life?* into What is "I"? According to the philosopher of science Henri Saget, life:

> *appears at times to reproduce the strict, causal sequences of a mechanism, at times to reflect the values of meaning and finality that belong to the mind.*[16]

He continues:

> *In a mathematical sense, the whole is the sum of the parts, but the tiniest organism is evidence enough that the whole is more than the sum of the parts.*

We may be critical of Schrödinger's reductionist approach but we cannot deny its heuristic value. It was adopted by the pioneers of molecular biology and led to spectacular progress partly because, when a reductionist approach encounters barriers or shortcomings, these can be identified. When we read Monod's comment that "*the structure and function of the chemical machinery is intrinsically the same from the bacterium to man*",[17] we know full well that the developing human embryo cannot be described entirely by mechanisms discovered in bacteria and that the description must include other kinds of knowledge. Nevertheless, didn't the study of nutrition in bacteria provide the fundamental data on the regulatory mechanisms of gene expression which form the basis of embryogenesis?

We should not exaggerate, however, the shortcomings of Schrödinger's approach. He himself stressed that it should not be taken as strictly positivist or essentialist. (We have called a view or representation of the world based on an array of types or essences "essentialist").[18] His heirs, especially molecular biologists, are perhaps less "flexible" than he was when they declare that there is a genetic order, or that every biological phenomenon depends strictly on the genetic message inscribed in DNA and transmitted to the cell machinery. Their view, inherited from essentialist approaches to life, confers DNA a sort of primordial invariant status by insisting on the self-reproducing and information-rich structure of the molecule. According to Ernst Mayr, DNA contains a programme, a determinant, a:

> *coded or prearranged information that controls a process (or behaviour) leading it toward a given end.* [19]

Whereas a machine is made, in a "*mortal duality*"[20], of matter coming from nature and a form imposed by man, living matter is, on the contrary, perfectly "informed" thanks to DNA. Its organisation does not come from without but from within; some authors speak of "pure form". The information in DNA is passed on by its replication, and it is because this is a permanent process that we can speak of order.

How do the protagonists of essentialism view evolution? World order is based on a limited number of essences or forms characterised by two main attributes, constancy and discontinuity. There is no room for fixism. Evolution can perfectly well occur discontinuously via *quanta* and display regularity. In the end, what is life but a well-ordered array of species and populations whose lack of homogeneity, if any, is resolved in a virtually unalterable genetic arrangement.

This essentialist view is rather hard pressed to make room for mutations. For Schrödinger, a very high mutation frequency would be an unbearable burden to living beings (see our discussion of the concept of load). That is why:

> *in order to be suitable material for the work of natural selection, mutations must be rare events, as they actually are. [...] The comparative conservatism which results from the high degree of permanence of the genes is essential.*[21]

On the other hand, Schrödinger suggested that we could view mutations positively as a means of introducing novelty. Novelty counterbalances the overall weakening of natural processes that occurs at each generation (if we accept that the overall impact of time is negative). His suggestion is, however, quite incidental. Genetic mutations (such as extinctions in palaeontology) are errors or stigmas with respect to a biological ideal, the ideal of the reproduction of an invariant order or of an organisation that should tend toward immutability.

In the end, it is best to consider a mutation as a chance or accidental event that modifies genetic material by conferring it a rare form. The new form can be integrated into the genome only with the help of, and not in opposition to, DNA-specific processes (in particular repair mechanisms which detect changes, try to remedy them, but sometimes fix them). Only the phenotypic expression of the mutation will tell us whether it is a deleterious or beneficial mutation, an error or not; no one will deny that mutations at the origin of cancers are bad.

The effectiveness of Schrödinger's essentialism depends partly on the constructivist method. Although it has its faults, it is flexible enough to avoid excessive reductionism. Living structures, at all levels, are not just machines obeying external orders but can organise themselves and reproduce. This aspect of Schrödinger's thinking certainly helped introduce the notion of information into biology.

## 4. AN UPDATE ON INFORMATION

In everyday speech, information is an item of knowledge, of scholarship, or an insight that the "speaker" has and that he (or she) can, if he wishes, either pass on to others or keep to himself. Information is a widely accepted notion and has a key role in our descriptions of power, of people's actions (how rational is their behaviour), of the way things are organised (data availability can help rationalise the way a complex system is set up and regulated), of automation and know-how. On the other hand, information theory is all about a statistical approach to communication. It is the outcome of a considerable amount of work on the optimal use of information transmission systems such as telephones, televisions, *etc*. The first person to provide an overview of information theory was Claude E. Shannon,[22] an engineer at Bell Telephone Laboratories, who showed that Boolean algebra, which reduces logical reasoning to an algebraic calculation in binary notation, can be translated into a physical system of open and closed circuits. The electric signal carries the information which is then transmitted via a channel (telephone line, Hertzian waves). We can thus study separately the actual information itself (amount, entropy of the source, *etc*.), the properties of the channels and, for optimal channel use, the relations between the information to be transmitted and the channel used. What the transmitted messages actually mean, in other words, the semantic aspect of communication is ignored. The basic concepts of this theory are simple and easy to generalise (information and amount of information, entropy, code efficiency and redundancy, coding in the presence or absence of noise, channel capacity, speed of transmission, *etc*.). It was therefore soon obvious that the concepts could be applied not only to the communication means of modern technology, but also to other fields, such as for instance, linguistics[23] or nervous system physiology. But does human communication really belong to this conceptual framework? Is it reasonable to reduce the whole world and its interconnections down to a set of elementary and distinct events that are translated into data linked by explicable rules? This is what scientists claim to be doing, at least in part, when they make use of the notion of information.[24]

In biology the notion of information has been used to address entropy (which, at a first approximation, measures the disorder of a system) and genetics. The idea that "information is nothing but negative entropy" is still controversial, resting as it does on rather improbable hypotheses. It turns living beings into creators of order in the Universe, and this view cannot be accepted without some discussion. Similar caution and doubts need to be voiced when we apply information theory to genetics. How can we transcribe the concept of memorising? Can we speak of innate and acquired information in genetics? How should we view the comparison made by some scientists between chromosomes and sources that emit signals and the idea that the sense organs pick up outside information, transmit it to the central nervous system, transform it, and then store it in genetic form? All these processes are indeed still somewhat hypothetical.

Clearly, biological information is not reducible to Shannon's model if only because the decoding systems are part of the information itself. Much progress is needed before information theory can account for genetic information as it has yet to define the meaning of an information and identify nonsense messages. It is a shame that mathematicians who are experts in information theory forget to mention these shortcomings and confuse the weaknesses of their constructions with a flaw in neo-Darwinian theory. We must therefore make quite clear what genetic information really means.

## 5. GENETIC INFORMATION

A willow tree, catkins, fluffs of cotton wool, the banks of a canal... all the features of a true English landscape are depicted by the eminent Oxford zoologist, Richard Dawkins, in his book *The Blind Watchmaker* when he introduces the theme of genetic information. But the poet soon gives way to the biologist. The willow's fluffy specks contain a portion of DNA:

> *whose coded characters spell out specific instructions for building willow trees that will shed a new generation of downy seeds. [...] It is raining instructions out there; it's raining programs; it's raining tree-growing, fluff-spreading, algorithms. That's not a metaphor, it is the plain truth. It couldn't be plainer if it were raining floppy discs.*[25]

It may seem natural to want to apply the concept of information to genetic material, given the concepts of primordial invariant, code, and messenger RNA, but this must be done with caution and due regard for aspects of life such as evolutionary pressures, mutation and selection.

1. We have seen that the way we perceive mutation depends in part on the importance we confer to the notions of normality and invariance. Is a mutation an "error" or just a change whose relevance we shall discover later? How can we reveal a mutation; by noting the impact on the phenotype, observing the chromosomes, or analysing the nucleic acid sequence? Moreover, which element in the information chain actually undergoes mutation; is it the source, the receptor, the transmission mode, or the information itself?

2. Analogous questions arise when we address selection. Two are virtually identical. Which entity is actually selected? At what integration level does selection occur? The second question was debated quite fiercely from the 1960s onwards although the protagonists themselves were not always fully aware that this was the question they were actually debating. V.C. Wynne-Edwards claimed that the group was selected whereas John Maynard-Smith and George C. Williams thought that selection occurred at the level of the individual or sometimes of the gene. As for the first question, Ernst Mayr defended the view that the selected entity was the individual

whereas Richard Dawkins, J. Maynard-Smith, G.C. Williams and William D. Hamilton said it was the gene. By drawing a distinction between the two questions and addressing the first one first, R. Dawkins may not have calmed the debate but at least he clarified it. According to Dawkins, we must distinguish between the entity subject to selection pressures (the individual) and that which is actually selected (the gene). Ernst Mayr accused him of turning Mendel's "wrinkled-pea" genetics into "beanbag" genetics! Chris J. Gliddon and Pierre-Henri Gouyon have classified Mayr's stance as "pea-soup" genetics.[26]

Because of mutation and selection, we must reformulate the issue of genetic information in a perhaps new, more fundamental fashion. We need to differentiate between what is transmitted over time and that which actually does the transmitting.

## 6. GENETIC INFORMATION AND AVATARS

Since that which is transmitted as faithfully as possible during evolution cannot reasonably be something ephemeral, what can it be? Is it a structure or is it information? Structure and information were no doubt synonymous in the first self-replicating systems but, with more elaborate organisms, the individual structure transmitted to the next generation became harder to define and the information easier. This explains why the idea of a gene has been so successful.

A gene is not easy to define as the number of definitions encountered in the literature prove:

> *Word invented by Wilhelm Johannsen (1911) to denote a material and self-catalytic particle which is harboured in the chromosome and which determines one or more inherited characters* (Grand Larousse Encyclopédique, 1962).

> *The characters developed by an organism depend upon the presence of genetic units which were initially called determinants and are now called genes. Genes have the power to reproduce as identical copies and are accurately transmitted during cell division* (Encyclopaedia Universalis, 1989).

> *Accurate copies in reproduction are guaranteed by genetic units that are fragments of chromosomes called genes. These genes pass on the characters of the parents during fertilisation: they are said to carry the genetic information* (Les Notions Philosophiques, 1990).

*Gene is a word that the Danish biologist W. Johanssen transposed into German in 1909 from the Greek Word genos meaning "birth, family, race", which can be related to the word gignestahai "to be born" and "to become" (since 1911, we also speak of "gene" in English). This biological term denotes an element of the chromosome that is responsible for a hereditary factor* (Dictionnaire Historique de la Langue Française, 1992).

*The chromosome theory of inheritance stipulates that each chromosome be made up of an ensemble of particles, each with its own role, and that are called genes [...] The gene's intrinsic property is the ability of self-reproduction: each time the cell divides, each gene of each chromosome gives rise to two genes that are faithful copies of it* (J. Rostand, L'Hérédité Humaine, 1952).

*The end term is the same irrespective of the type of analysis we use: the gene represents simultaneously the unit of function, mutation and recombination. The material of heredity is resolved into elementary units that cannot be split. The gene becomes a kind of atom of heredity.* (F. Jacob, La Logique du Vivant, 1970).

*We define the gene as the smallest segment of the DNA macromolecule that is able to determine a constant character* (P.-P. Grassé, L'Evolution du Vivant, 1973).

*Unit of inheritance, borne by a chromosome, transmitted from generation to generation by the gametes and controlling the development and characteristics of an individual* (translated from the French) (E. Mayr, Populations, Species, and Evolution, 1970 (see Glossary)).

*At some point a particularly remarkable molecule was formed by accident. We will call it the Replicator. It may not necessarily have been the biggest or the most complex molecule around, but it had the extraordinary property of being able to create copies of itself... They have come a long way those replicators. Now they go by the name of genes, and we are their survival machines.* (R. Dawkins, The Selfish Gene, 1976).

Several of these definitions describe the gene as a sequence of nucleic acids with defined properties. The gene is thus apparently a physical entity. But what should we make of gene cloning where a gene is reproduced in several copies? Do we then refer to several genes? No, we do not; the gene is unique. We should therefore not view a gene from a physical "nucleotide" stance but rather consider it as genetic information determining the observed phenotype. It is fundamental to note that: "That which is transmitted during evolution, whatever we call it, is not so much a physical entity as a piece of information (in the broadest meaning of the word that includes the notion of structure)." We should not confuse information and matter even if researchers in the West often find it hard to think in terms other than physical objects.

> *When an abstraction, such as the gene, suddenly comes on stage, the biologist will go to no end to replace it by physical elements, particles or molecules. It is as if a theory in biology could last only if it refers to some concrete model.*[27]

Biologists are not being urged to take part in a Copernican revolution, nor to convert to some kind of oriental spirituality. In any case, there is no need for this since the gospel according to St. John states that:

> *in the beginning was the Word... All things were made by him; and without him was not anything made that was made. In him was life... And the Word was made flesh.*[28]

They are merely being asked to include the modern notion of information in their thinking.

To denote that which transmits genetic information or its physical carrier, we use the term avatar borrowed from the Hindu religion; it alludes to the physical forms adopted by the god Vishnu on his visits to Earth. The word was introduced by John Damuth of the University of California to refer to the selection unit in the case of several competing species interacting in an ecologic community. An avatar is:

> *the local "embodiment" or "representation" of the species in the local community. A community is a population of avatars, each avatar belonging to a different species. [...] "Avatar" is a purely ecological concept, and has no relationship to taxonomic concepts such as "subspecies" or "race".* [29]

The avatar, as noted by J. Damuth, interacts with the environment which provides for its needs and exerts an influence upon it but, above all, the avatar is produced by genetic information to ensure that this information is passed on. Individual organisms easily meet this definition. They interact with the environment, are produced by genetic information, and copy the information. The number of copies depends on reproductive capacity, interaction with other avatars and with the environment, unforeseen events, and other factors. What is true for the individual must be extended to all of life's integration levels. Interestingly, with the exception of the word gene, the words most often used in biology refer to avatars and not to the corresponding information!

The distinction between that which is transmitted and that which transmits, between genetic information and its avatars, can help resolve the problem of which entities undergo selection. Selection targets only genetic information, avatars are mere vehicles.

Our conclusion on the gynodioecia of thyme - namely, that the hermaphrodite and female forms are not the units of selection - can be extended to the entire living world. Genes, not individuals, are selected, and this explains life's inventiveness. Genetic information, which is selected on the basis of an avatar's ability to reproduce this information, will produce living structures that are increasingly complex.

## 7. INDIVIDUALS: GENE INVENTIONS FOR GENE REPRODUCTION

Darwinian and neo-Darwinian theory have accustomed us to focus on the individual. We have just shattered the individual's privileged position of monopoly, and guided readers toward organisms, organs, and cells, right to the heart of living matter where genetic information lies. Some readers will tax us with reductionism, quite aptly at this stage because, as we have shown and will show yet again, a reductionist phase is often necessary to understand life's phenomena. We shall now take a wide-angle lens and look at the levels of integration mentioned above in another way.

Our definition of genetic information may have surprised readers more familiar with matter ("substance") than form ("logos"). Let us go one step further and qualify genes by an adjective used to describe humans (and sometimes, by analogy, animals); let us call them selfish and state clearly the central theme of this chapter: Individuals are artifices that genes have invented for the purposes of their own reproduction. This assertion is analogous to Richard Dawkins' statement on the human species:

> *We are survival machines - robot vehicles blindly programmed to preserve the selfish molecules known as genes.*[30]

Dawkins speaks of "molecules" whereas we and C.J. Gliddon speak of information. Whichever the word, the statement surprises, even shocks. But is it not just an extension of our thoughts on genetic information? Why should we limit our investigations to entities, "avatars", that will disappear, however complex, evolved and noteworthy they may be? Only genetic information will persist even though (or because) it is subject to selection and variation.

Once we have accepted R. Dawkins' broad statement, of what use is it to us? It will be our guide on our tour of the edifice of biology. This tour will take us from genomic events to the behaviour of populations and from the mechanisms of reproduction to ageing and death.

**Is reproduction a trap?**

*It is commonly said that living beings multiply. How untrue! Are we gullible or just naive? Individuals do not multiply, their death is programmed and they will disappear forever. If they have offspring, their children will never be "them", they will be "others", genetically other if they are sexual species (the offspring has neither parent's genotype) or just different even if they have been cloned (e.g. true twins because otherness exists even in twins). We are all aware of this but sometimes tend to forget it. No one has seen parents express horror and surprise because they have acquired sudden proof that their child is irrevocably another being who cannot be taken back.*[31]

When we say that a successful gene's prime quality is its selfishness which is expressed in an individual's behaviour, are we not undermining ethics, "*that set of rules of behaviour that are accepted unconditionally and that are considered to apply*"? Is it not pointless to fight man's selfishness if it is genetically determined? R. Dawkins writes discerningly:

> *Be warned that if you wish, as I do, to build a society in which individuals cooperate generously and unselfishly towards a common good, you can expect little help from biological nature. Let us try to teach generosity and altruism, because we are born selfish. Let us understand what our selfish genes are up to, because we may then at least have the chance to upset their designs, something that no other species has ever aspired to.*[32]

In other words, we do not have to obey our genes (an example is contraceptive use). Although the gene is fundamental to the way we view evolution, it is not the only basis of evolution, especially human evolution.

## 8. SELECTION LEVELS

Selection acts on genetic information at all of life's integration levels but in different ways at each level. Each selection level is a more or less independent unit.

## Molecular level

The fairly recent discovery that much of the genome is made up of repeat non-coding sequences that do not correspond to genes was a surprise to molecular biologists. Their immediate reaction, governed as usual by a physiological and invariably utilitarian reflex, was to find out of what use they might be to genome and organism. But here lies a misunderstanding we have already come across. The repeat sequences are not like simple organs that cannot reproduce on their own and that have to serve a purpose to remain part of a larger structure. These sequences can replicate independently of others and "invade" part of the genome. This idea underlies the selfish DNA hypothesis proposed by Leslie Orgel and Francis Crick in 1980 in an article that made the front page of *Nature*, "*Selfish DNA: the ultimate parasite*".[33]

Many molecular mechanisms covered by the generic term "molecular drive" can increase the number of copies of a sequence in a genome: Gene conversion which promotes homogeneous sequences (two homologous but non-identical sequences pair up to form two heteroduplexes; DNA repair enzymes then resolve the differences to the advantage of one of the forms and yield two identical sequences), asymmetric crossing over, "sliding" (the polymerase responsible for replication "stutters" and yields a series of identical copies), transposition, *etc*. These "parasitic" processes at the molecular level can explain repeat sequences as well as phenomena of concerted evolution;[34] they might even be beneficial to avatars, whether these are individuals or populations. According to G.A. Dover, molecular drive might play a role in most, even all, evolutionary events.

Whatever the outcome of these molecular mechanisms at the individual level, they are clearly important early on in the evolution of living beings. Thus, introns (see Chapter 3), which promote crossing-over by increasing the chances of recombination, seem to have been present before prokaryotes and eukaryotes separated. Transposition mechanisms, in which RNA may be an intermediate, have probably existed for a very long time. The enzyme required, reverse transcriptase (which catalyses the transition from RNA to DNA unlike transcriptase which catalyses the transition from DNA to RNA), might date back to an RNA world that existed before our DNA world. The claim that RNA preceded DNA, and that reverse transcription preceded transcription, is not that outrageous if we recall that organisms and their genomes (in a physical sense) are the means or avatars that genetic information units have invented to maximise their multiplication over time.

## Cellular level

The next level is the cellular level. A bacterium's only dream is to become two bacteria, wrote F. Jacob. Cell multiplication in higher organisms, however, is usually governed by a development plan. At times, it might be worth escaping from the grips of this plan. Thus, cancer cells, because they divide more often, have a selective advantage over an organism's other cells. In animals at least, there are mechanisms for eliminating such dividing cells before their "victory" ends in the death of the individual. Oncogenes, whose avatars are cancer cells, do not always have a winning bet because, for instance, some anticancer drugs act by creating a selection pressure against cells that divide frequently.

## Individual level

The level of the individual is the one that has received most attention since Darwin's day. At this level, the selection of genetic information depends directly on the avatars' ability to reproduce. However, in the case of sexual reproduction, there is the problem that only part of the genetic information is passed on. Ensuring that an avatar reproduces successfully does not guarantee that the information will be passed on to the next generation. Each diploid individual only passes on one chromosome of each pair in a random combination. Moreover, because of recombinations within each pair, the assorted pieces of information on the inherited chromosome are also random. Hence George C. Williams' query, what is the smallest amount of information that can have an effect on an individual's phenotype? The greater the amount of information, the larger the space it occupies on the genome, and the smaller its chance of being transmitted in its entirety. In point of fact, the selected information can be as small as a single nucleotide (that had for instance undergone mutation) or as large as a haploid genome (many cases of polyploid selection are known in plants). The information in a nucleotide has a 50 percent chance of being transmitted. Moreover, the meaning of this information may depend upon the rest of the genome which is usually not transmitted in its entirety. We have yet again to define an individual but this time scientifically. The information carried by the avatar/individual is not a single unit; each piece of information is selected according to its own logic.

The question "which is the selection unit" has special prominence at the level of the individual. How can we specify the exact relationship linking a piece of genetic information to its avatar? E. Mayr considers the individual as a whole, as a carrier of a complete genome. R. Dawkins and G.C. Williams, as we have seen, stick to the gene. Nevertheless, just as in sexual species a genome is never passed on in its entirety, a gene is never passed on in isolation. The selection unit would thus be a set of genes. Since the variation in the frequency of an allele depends on the reproductivity of all its carriers and since in each carrier the allele is in a genetically different environment, the allele's fitness is clearly a statistical, and not an intrinsic, property.

## Group level

V.C. Wynne-Edwards introduced the concept of group selection in 1962. He used it to explain so-called altruist behaviour in animals. For instance, whereas most prairie dogs go about their usual business, some look out for the unexpected arrival of predators. The lookout informs the rest of the group of the predator's presence by sending out a sound signal but, in so doing, runs the danger of attracting the attention of the predator. How did lookout genes become fixed if the lookout dies before passing them on? Many types of insect spend their lives looking after the offspring of their sisters at the expense of their own. Similar behaviour is observed in mole-rats. How can we explain a behavioural pattern that natural selection should have eliminated long ago? V.C. Wynne-Edwards suggests that the prejudice to the individual is offset by the benefit to the group, which is thus the selection unit. But this is not the only possible hypothesis.

*Figure 4.4.* Prairie-dog on the look-out for the arrival of predators. The animal informs the rest of the group of the predator's presence by sending out a sound signal but in so doing, runs the danger of attracting the attention of the predator. How did look-out genes become fixed if the look-out dies before passing them on?

William D. Hamilton has suggested the following explanation. Unselfish behaviour is always encountered in closely related individuals. In other words, the gene that "commands" unselfish behaviour is advantaged because it is also present in the individual who benefits from the altruism. The same reasoning also applies, for example, to the caring behaviour of parents towards their children. The gene acts for its own benefit because it is present in both generations. W. Hamilton also explains how unselfish behaviour is maintained and even developed in some species outside the strict bonds of lineage. In bees, because of their special haplo-diploid mode of sexual reproduction, a female has more genetic points in common with her sister than with her daughter. This is why workers look after their sisters rather than produce offspring. At the root of the expression, selfish gene, devised by R. Dawkins, is the idea that the help one individual gives to another conceals a tendency to maximise the reproductive worth of the gene.

Hamilton's hypothesis, baptised "kin selection" by J. Maynard-Smith, underlies a good predictive model. In contrast, Wynne-Edwards' view (a character, even if it is detrimental to the carrier, will be inscribed in the species if it is beneficial to the group) is hard to defend. Populations do not display sufficient individuality to be true selection levels.

We must therefore probably abandon the idea that some hypothetical group selection might explain the decrease in an individual's fitness due to unselfish behaviour. As a first approximation, individual selection alone can explain group evolution and, when it seems to be absent, we should not immediately invoke group selection. Nevertheless, during evolution, groups of different sizes (species, phyla) have gone extinct, and current groups are probably not random samples of what might have existed. Preferential group extinction which is a kind of group selection (we have rejected it as an explanation for decreased individual fitness) has therefore no doubt reduced the number of situations where individual selection could act (we shall come back to this when discussing the evolution of sex). Living species are thus clearly not random samples taken from among the results of individual selection. It is unfortunately difficult to evaluate the true impact of this phenomenon. How can we measure group fitness? On the basis of what property? We lack answers to these two questions.

## 9. RULES GOVERNING THE EVOLUTION OF INFORMATION

Genetic information reproduces by using avatars as carriers and as procreators of other avatars meant to carry identical information. In practice, avatars with finite resources cannot provide exact copies of the information they carry over time. There is a minimum, inevitable amount of diversity for a variety of reasons. Consequently, different information is reproduced by avatars using the same resources. There is competition, and the medium is invaded by the information that is reproduced best, whether by chance (or genetic drift) because it produces the best avatar (the selection process *per se*) or because it is linked to other information that is selected (hitch-hiking).

*Table 4.1.* Life, the universe and the rest ... This table shows the different kinds of information and their avatars. At each integration level (not all are listed), information is carried by a physical entity whose only raison d'être is to carry this information. It is merely the avatar of the information. All living things carry information and are the products of the evolution of this information. When a new piece of information emerges, it first crops up in a nucleic acid molecule. It can then affect the chromosome, cell, and carrier and even influence the fate of the population and species. It will be selected at all these levels in the order of priority indicated by the direction of the arrow. Information counterselected at the cell level will never be carried by an individual ... The way it all works is even more complicated than this because each level arising from all previous levels makes up the environment (see direction of arrow indicating environment). (Adapted from Gouyon, P.H. and Gliddon, C., The genetics of information and the evolution of avatars[35]).

| Information | Avatar | Science | Mechanism | |
|---|---|---|---|---|
| 1 sequence | several nucleotides | | Molecular drive | |
| | | Molecular | | Environment ↑ |
| 1 gene | several nucleotides | Biology | Mutation | |
| 1 set of genes | 1 chromosome | | Recombination | |
| 1 genome | 1 cell or virus | Population Genetics | Selection | |
| 1 set of n genomes | 1 individual | | Drift | |
| 1 gene pool | 1 population | | Migration | |
| 1 wider gene pool | 1 metapopulation | | Dispersal<br>Speciation | |
| 1 set of gene pools | 1 species | | Competition | |
| | | Ecology | Predation | |
| 1 set of sets of gene pools | 1 ecosystem | | Symbiosis<br>Succession | |
| | | | | Priority ↓ |
| The Word? | Saint John? | | | |

This relatively simple process is in fact more complicated because there is not just a single unit or level of selection. Several levels have to be considered simultaneously and in ranked order. We have already encountered this in group selection but the rule can be generalised: All forms, even weak forms, of selection at a given level will dominate over selection at higher levels even if this "conflict" may lead to extinction ... Thus, for a character to be selected at the individual level, it has already to have been selected at the molecular level. Similarly, only the characters that are successful at the level of individual selection can hope to be selected at group level (Table 4.1).

We should interpret this hierarchy in a circular manner rather than in the usual linear way. A well-known example is that of the wings of birds. Do birds have wings because they fly or do they fly because they have wings? Owing to mutation and development, which act at the molecular and individual levels, birds fly because they have wings. Owing to selection, which acts at the individual and population levels, birds have wings because they fly. The circular nature of the causes and the hierarchy in life's integration and selection levels hinge on each other; they let us reintroduce the notion of finality in its rightful place, to the right extent.

The hierarchy in levels also means that efficient reproduction at one level does not imply efficient reproduction at other levels. If a piece of genetic information corresponding to a given nucleotide sequence is reproduced correctly by the enzyme machinery, it can increase its number of copies in the genome... but these extra sequences will probably not benefit the carrier. We have already mentioned the higher level case of cancer cells. There are many other examples.

The "unreasonable" proliferation of cancer cells (they will die with the patient) is an incentive to examine situations where several finalities co-exist at the same integration level. What will happen? Alas, for those who admire natural harmony or hanker after the Garden of Eden, the result will be conflict. The conventional view of the genome is thus somewhat modified; it becomes a place of perpetual conflict.

## NOTES

1. Charnov, E.L., Maynard Smith, J., and Bull, J.-J., 1976, Why be an hermaphrodite? *Nature* 263: 125-126.
2. This rather curious terminology refers to the fact that the female flower is a hermaphrodite flower that has lost its male function.
3. Östergren, G., 1945, Parasitic nature of extra fragment chromosomes, *Botaniska Notiser* 2: 157-163.
4. Two lawyers in England based the concept of a pre-embryo on indivisibility. The pre-embryo is a small group of cells that make up an embryo of less than 14 days. It can be used in experimentation because it is not yet an individual and can divide into twins. We lack the space to discuss the ethical issues involved.
5. Schrödinger, E., 1944, *What is Life? The Physical Aspect of the Living Cell*, Cambridge University Press, Cambridge.
6. Schrödinger, E., *ibid.*, p. 1.
7. Schrödinger, E., *ibid.*, p. 2.
8. Schrödinger, E., *ibid.*, pp. 68 & 69.
9. Haldane, J.B.S., 1931, *The Philosophical Basis of Biology*, London, p. 2.

10. Bohr, N., 1933, Light and life, *Nature* 131: 457 & 458.
11. Quoted by E. Schrödinger, 1954, in *Nature and The Greeks,* Cambridge University Press, Cambridge.
12. Monod, J., 1970, *Le Hasard et la Nécessité. Essai sur la Philosophie Naturelle de la Biologie Moderne*, Seuil, Paris, p. 118.
13. Schrödinger, E., *What is Life*? pp. 19-20.
14. Jacob, F., 1970, *La Logique du Vivant. Une Histoire de l'Hérédité*, Series: Bibliothèque des Sciences Humaines, NRF-Gallimard, Paris, p. 272.
15. Schrödinger, E., *ibid.*, p. 30. A gene, in fact, contains several tens of thousands of atoms, those of the nucleic acids.
16. Saget, H., 1978, L'essor de la biologie moléculaire (1950-1965), *Cahier d'Histoire et de Philosophie des Sciences*, n° 7, CNRS, Paris, p. IV.
17. Monod, J., *ibid.*, p. 118.
18. See Chapter 1, Section 1.
19. Mayr, E., 1981, *La Biologie de l'Evolution*, Hermann, Paris, p. 117.
20. Saget, H., *ibid.*, p. VI.
21. Schrödinger, E., *ibid.*, p. 41.
22. Shannon, C.E., 1948, A mathematical theory of communication, *Bell System Techn. J.* 27, 379-423 & 623-656 and Communication in the presence of noise, 1949, *Proc. Inst. Radio Engin.*, 37: 10.
23. In linguistics, the channel is the chain formed by the organ that produces sounds, the sound waves themselves, and the organ that hears the sounds.
24. Brillouin, L., 1959, *La Science et la Théorie de l'Information*, Masson, Paris.
25. Dawkins, R., 1986, *The Blind Watchmaker*, Longmans, London. (Penguin Books, 1991, p. 111).
26. Gliddon, C.J. and Gouyon, P.H., 1989, The units of selection, *TREE 4*: 204.
27. Jacob, F., *La Logique du Vivant*, p. 22.
28. Testament of St. John, Chapter 1.
29. Damuth, J., 1985, Selection among "species": A formulation in terms of natural functional units, *Evolution* 39: 1137.
30. Dawkins, R., 1976, *The Selfish Gene*, Oxford University Press, Oxford, preface (new edition 1989).
31. Gouyon, P.-H., 1994, Le sexe, ce choix de l'évolution biologique, *Turbulences* 1: 9.
32. Dawkins, R., *ibid.*, p. 3.
33. Orgel, L.E., and Crick, F.H., 1980, Selfish DNA: The ultimate parasite, *Nature* 284: 604-607.
34. Sequences often do not evolve independently of each other.
35. Gouyon, P.H., and Gliddon, C., 1988, The genetics of information and the evolution of avatars, in: *Population Genetics and Evolution* (G. de Jong, ed.), Springer, Berlin, p. 121.

# Chapter 5

# CONFLICT AND COOPERATION

It is hard to imagine that components of an organism such as the cytoplasmic and nuclear genomes of thyme described in the previous chapter may enter into conflict. Do not different genome compartments[1] share the same evolutionary fate? Are they not transmitted together to the next generation and selected so that the carrier reproduces as successfully as possible? By underscoring genetic information's fundamental role in the new theory of evolution, the previous chapter has given a hard knock to the conventional harmonious view of the "individual". It has called into question whether a gene's (or group of genes) evolutionary success does really depend on how useful the gene is to the individual. This chapter will confirm our doubts and show that competition - intragenomic conflict - does occur in individuals. It will then present two theories, game theory and sociobiology, in order to explain how conflicts are managed at different levels in living beings.

## 1. THE GENOME, AN ARRAY OF COMPARTMENTS IN CONSTANT CONFLICT

According to the previous chapter, the genome is made up of pieces of information that interact to "manufacture" an individual that will pass them on. However, interaction does not imply coherence. The genome is not a homogenous entity but formed of several compartments each of which has its own mode of operation and, especially, reproduction. As we shall see below, transposons - baffling repeat sequences discovered in the 1970s - can give rise to several subcompartments within the genome. We have already learnt that the nuclear and cytoplasmic genomes of thyme form two quite distinct and probably ranked compartments.

How do these different compartments behave toward each other? Traditionalists were quick to spy instances of co-operation or independence. However the genome bears many traces of conflict. Sequences that were "vanquished" in the past, and whose multiplication machinery is inhibited for the time being, are always on the ready to be expressed again. A single mutation or interbreeding could re-establish contact and interaction between pieces of information that lie apart (or separate two antagonist pieces of information), and set up the sequences so that all restrictions are lifted.

## Conflict between asexual chromosomes

Meiosis is cell division in sexually reproducing species. There is passage from the diploid state (where the individual is endowed with two sets of chromosomes, one inherited from the father, the other from the mother) to the haploid state (the sex cells, or gametes, which have one set of chromosomes only and therefore just one copy of each gene). Meoisis ensures that genetic information is distributed equitably; half of the gametes (spermatozoa or ovules) receive one copy of the gene, the other half receives the other copy. Equitable distribution of alleles to gametes is a conclusion of Mendel's work. That it does not always occur implies that meiosis is a time of the life cycle that is especially prone to conflict.

Let us consider a heterozygote with, by definition, two different copies of a given gene. These two allelic forms compete with each other for presence in the gametes. If one can bias meiotic segregation so that it will be present in most gametes, then this "trait" will be strongly selected. This is known as meiotic distortion, meiotic drive or segregation distortion and can lead to the disappearance of the other allelic form.

The segregation distortion factor ensures that one allelic form is often over-represented by eliminating some or all of the gametes where it is not present. This form will be selected if the number of gametes is not limiting, that is if destroying half of them does not decrease fertility. This is why segregation distortion is usually observed in males. An example is meiotic distortion due to the special allele, haplotype t, in the mouse. Because, after meiosis in a heterozygote t/+ male, haplotype t is absent in half of the gametes, cell viability is decreased.[2] Haplotype t is thus at a clear advantage in the progeny of heterozygotes because, instead of being present in just half of the gametes (as per Mendel's laws), it is sure to be present in nearly all surviving spermatozoa.

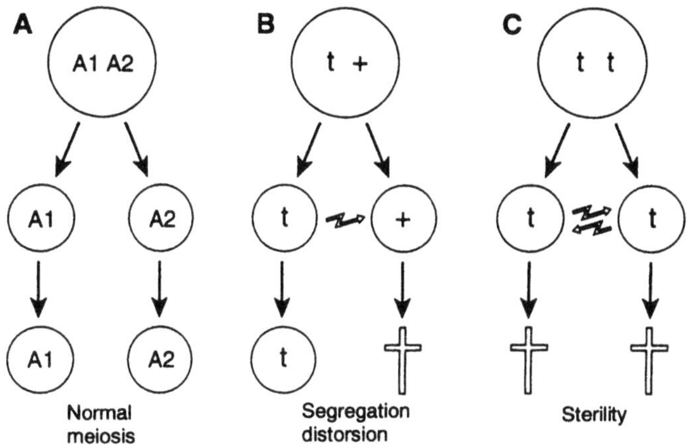


*Figure 5.1.* During meiosis, genetic information is usually distributed fairly between the gametes. For example, an A1/A2 heterozygote will provide one half of A1 gametes and one half of A2 gametes (left). Haplotype t inactivates the spermatozoa in which it is not present after meiosis by a poison produced before meiosis. The spermatozoa in which haplotype t is present are resistant to this poison. It thus makes sure that it will be in all functional spermatozoa produced by a t/+ heterozygote (middle). This is called meiotic segregation distortion. Most often, however, it kills off t/t homozygotes or, if they are viable, renders them sterile (right). (Adapted from Atlan, A. and Gouyon, P.H., 1994, Les conflits intragénomiques, Société française de Génétique, pp. 1-9).

If there are two copies of the allele however, the homozygote males are sterile or die, without us knowing whether the damage was due to mutual destruction of gametes (explaining sterility) or to lethal recessive genes near the locus of the distortion factor (explaining death). Because the presence of haplotypes inducing sterility or death in the homozygote state is detrimental to the entire genome, these genes are usually eliminated rapidly by natural selection. Haplotype t has avoided this fate because the reproductive advantage endowed by segregation distortion keeps it in 10 to 20 percent of the natural population. The conflict between haplotype t and the remainder of the mouse genome can only be resolved by eliminating the haplotype or its adverse effects.

Meiotic distortion factors are found in many species. Those that can be observed - that enter into conflict with the genome and thus do not become fixed - have a more or less detrimental influence on the remainder of the genome. When there is no adverse effect in the homozygote state and no conflict, the distortion factor rapidly invades the population. Once it is fixed, its effect is undetectable because it is inherited by all gametes. Our genome is probably riddled with old distortion factors that are without effect because present in all of us. A special case is the presence of distortion factors in the sex chromosomes.

## Conflict between sex chromosomes

In many species, gender is determined by sex chromosomes. One of the sexes is homogamous (it has two homologous sex chromosomes); the other is heterogamous (it has two different sex chromosomes). Consequently, the heterogamous individual determines the gender of the offspring. Let's take the example of mammals where the female is homogamous (XX) and the male heterogamous (XY). When meiosis proceeds correctly, half of the male gametes carry an X chromosome and the other half a Y chromosome, thus ensuring equal numbers of male and female offspring. However, a chromosome that can bias segregation so as to predominate in the gametes will be selected. This alters the sex ratio of the offspring. If, for example, the X chromosome biases meiotic segregation so as to be present in virtually all spermatozoa, nearly all of the offspring will be female.

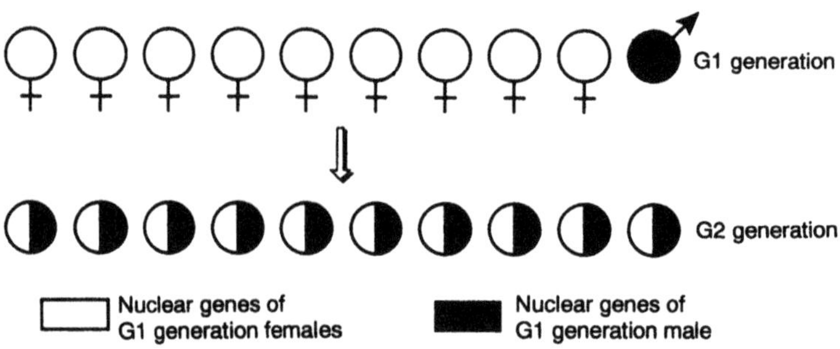


*Figure 5.2.* The reproductive advantage of the rarer sex. Let us consider a population of 10 individuals of unchanging number from one generation to the next. If the G1 generation comprises 9 females and one male, this male will sire 10 G2 offspring (since he is the only father available) whereas each female will have an average of 1.1 G2 offspring. The nuclear genes of the G1 male will therefore be passed on better than the genes of each female. (Adapted from Atlan, A. and Gouyon, P.H., 1994, Les conflits intragénomiques, Société française de Génétique, pp. 1-9).

If there were just meiosis, there would be equal numbers of males and females in a population. However, meiotic distortion of sex-determinants opens the way to a wide range of sex-ratios. It is therefore surprising that, irrespective of whether genetics or environment determine sex, we nearly always encounter equal proportions of males and females. Why? Would we not expect the optimal sex ratio to be rather low and biased against males as we only really require the number of males able to fertilise all females? In other words, for one and the same number of offspring, it does not matter whether there are 50 percent of males and 50 percent of females, or 99 percent of females and 1 percent of males.

In 1930, Ronald A. Fisher was the first to realise that, in order to explain the male-female equality in numbers, we should not focus on the number of children but grandchildren. In 1967, W. Hamilton put Fisher's argument as follows: Imagine a population with more females than males at birth. In such a population, a young male has more chances of mating than a female and will, on average, have more offspring. Parents genetically disposed toward having more males than females will have more grandchildren. Genes determining male gender will thus increase in frequency until we reach a balance between male and female production and the advantage of producing more males is reduced to zero. The reasoning works whether we start off with more males or more females and is summarised in the statement the rarer sex is the one whose representatives transmit the most genes.

Thus a distortion factor cannot invade an entire population as this population would end up with individuals of just one sex and become extinct. In fact, the trend is interrupted sooner or later because the minority sex has more leeway to expand. Because it cannot become fixed, a distortion factor carried by a sex chromosome, unlike one carried by a non sex chromosome, creates a conflict between itself and the entire nuclear genome. Notwithstanding, many instances of sex segregation have been observed in natural populations. The one that has attracted most attention is the X chromosome of the fruit-fly. The genes that repress this distortion are located throughout the entire nuclear genome.[3]

## Conflict between nuclear and cytoplasmic genomes

We have seen that self-fertilisation in hermaphrodite plant species ends up by reducing the "cost" of producing males. Once we accept that evolutionary forces exert different actions on different genome compartments and that it is possible to redistribute the resources of an impeded male route, we can understand that it is in the interest of cytoplasmic genes, which are passed on only by the female, to hold the male route in check. For example, cytoplasmic genes in hermaphrodite species have been selected to prevent pollen production. Nuclear genes which are transmitted by both males and females are a different case. The greater the number of male gametes in which a nuclear gene is present, the more copies it will leave. Selection's action on cytoplasmic genes is offset by its opposite action on nuclear genes. Ever since male sterility has existed, nuclear genes have been selected in order to restore male fertility to sterile-male cytoplasms. We explained this mechanism at length in Chapter 4 using thyme as an example.

Several studies have modelled the conflict between nucleus and cytoplasm and shown that it will maintain polymorphism in the population if we make the following hypothesis. If the population is not to be invaded by the allele that restores male fertility, we must assume that, in the absence of the cytoplasm being restored, this allele is at a disadvantage compared to non-restoring and non-maintaining allele(s). This is probably true if we consider the sterile cytoplasm as a kind of "internal pathogen" and the restoring allele as a resistance gene. There are many published examples of resistance genes that, in the absence of pathogen (or pesticide), are at a disadvantage compared to the "sensitive" allele. In the area around Montpellier in France, mosquito populations that had become resistant to DDT became extinct less than 10 years after DDT use had been discontinued.

It is therefore quite reasonable to assume that natural selection in hermaphrodite populations has given the edge to cytoplasmic information that impeded male gamete production. By transferring the spared resources to females who alone pass this information on, it enhances their reproductive fitness. This cytoplasmic information has appeared in many plant species but has been virtually silenced in contemporary species by subsequent nuclear restoration genes. Cytoplasmic genes may, at times, wield "power" over the respiratory chain (some of whose proteins are encoded by the mitochondrial genome) by blocking androgenesis. During their history, they have devised several ways of blocking androgenesis and, for some, this is their sole function. Male sterility cannot, however, be reduced to a mere question of cell respiration. Its main cause is evolutionary rather than physiological.

The endless race between cytoplasmic sterilisation and nuclear restoration can take place on many playing fields. If it goes on independently in two species that diverged some time ago, the cytoplasm of one species combined with the nucleus of another not unsurprisingly causes male sterility rather than any other malfunction because the restoring nuclear information is inadequate. In general, the male function can be easily perturbed by events such as tissue culture, thermal stress, transposition, *etc.* The network of interactions between nuclear and cytoplasmic information is the result of constant tinkering ("*bricolage*", to quote François Jacob) and integrates all past events. Odd interactions and between-compartment gene transfers occur that make the ensemble fragile.

## Conflict between the nuclear genome and endosymbiotic parasites

Many examples of "selfish cytoplasms" in animals are not due to mitochondrial genes but to endosymbiotic parasites. Several species of arthropods harbour endosymbiotic bacteria in their cell cytoplasm. Like mitochondria, these bacteria are inherited maternally and undergo the same evolutionary pressures as outlined above. The host species are not hermaphrodites but males and females. Because males do not pass the bacteria on, any gene that makes sure that the bacteria are not present in males or that sacrifices males for the benefit of females is selected.

Many natural processes let endosymbiotic bacteria multiply at the expense of the nuclear genome. The most spectacular case is probably that of the wood-louse *Armadillidium vulgare.* The bacterium that infests a genetic male wood-louse transforms it into a functional female, thus substantially increasing its own reproduction. Instead of being a male that will not transmit the bacterium to any of its offspring, the wood-louse is a female that will pass it on to nearly all its progeny. On the other hand, multiplication of the nuclear genome of the parasite-infested louse is decreased because this would boost the number of females. As we have already mentioned, the nuclear genes borne by the predominant sex in a population are disadvantaged whereas the genes counteracting this dominance are advantaged. This explains why nuclear resistance genes against feminisation are found in populations of *Armadillium vulgare.*

Similar events occur in other species. In Trichograms (the insects, hymenopterans), bacteria endow females with the ability to reproduce clonally, by parthogenesis, thus causing extinction of males. In ladybirds, bacteria kill off their male hosts. They gain a reproductive advantage because resources are made available to their female hosts who thus display improved survival.

Are these valid examples? The conflict arises in one and the same individual but relates to organisms from different species. Admittedly, but is not the distinction between endosymbiotic bacteria and mitochondria rather artificial? Many biologists, like for instance Lynn Margulis, consider that current organelles arose by endosymbiosis between the ancestors of eukaryote and prokaryote cells. According to serial endosymbiosis theory, the ancestors of mitochondria were aerobic bacteria which appeared when oxygen spread over the surface of the Earth. By absorbing aerobic cells, anaerobic cells could use oxygen and thus become aerobic also. The crucial question is how this kind of symbiosis became inheritable and a potentially powerful evolutionary driving force. L. Margulis speaks of genomic predation and has suggested that eukaryote cells may have acquired a wide range of functions by absorbing the genomes of other organisms together with their "findings" (such as flagella, respiration, photosynthesis, *etc*).

Whatever actually happened, we are dealing here with a key issue in the study of genomic conflicts. What are the true boundaries of an individual's genome? The boundaries of conflict are even less well delimited in other examples.

## A particularly virulent chromosome

In the 1930s, an extra chromosome, called B chromosome, was discovered first in plants, then in other organisms. Surprisingly, although a disproportionate number of B chromosomes are present in gametes and this number increases with each generation, individuals do not seem to have many B chromosomes. These chromosomes have apparently reached a state of equilibrium in natural populations. Biologists immediately asked of what use they might be to the carrier. They could not answer the question because individuals lacking a B chromosome did not seem to be disadvantaged in any way. It was not until 1945 that Östergren postulated that the extra chromosomes were just parasites. At the time, the idea was quite revolutionary but it is now part of the more general notion of a selfish replicator. Although the presence of the extra chromosomes is a sign that they are useful, "*they need not be useful for the plants. They need only be useful to themselves*".[4]

The presence of B chromosomes can have an adverse effect on an individual's genome. An extreme case is the paternal sex ratio (PSR) chromosome which, when it manages to enter a genome, writes off all other chromosomes. It has been detected in the hymenopteran *Nasonia vitripennis* through its influence on the sex ratio. In Hymenoptera (bees, wasps, ants), the sex of the descendants is determined by the mother during egg laying. If she fertilises the egg with spermatozoa in her spermatheca, the descendant is diploid and female; if she does not, the descendant is haploid and male. This system means that males normally inherit nothing from their "father" (their mother's male partner).

Male PSR chromosome carriers have been identified by the high proportion of males among their progeny. This high proportion could simply be due to partly sterile sperm (ill-fertilised eggs would just contain the maternal genome) but these male offspring can, just like their fathers, distort the sex ratio toward an excess of males. What do they inherit from their fathers? The PSR chromosome, of course. Present in the sperm this chromosome inactivates all other chromosomes by their condensation into heterochromatin during fertilisation. They are lost during the first meioses so that the PSR chromosome soon becomes the sole surviving paternal chromosome. In the end, the embryo develops as a male that contains only maternal chromosomes together with the PSR chromosome. The fate of the maternal chromosomes, however, is also sealed because PSR will eliminate them when they turn up in sperm. In summary, PSR reproduces well from father to son but no chromosome with which it has been in contact will be passed on to the next generation.

## Conflict between paternal and maternal genes

When the mother provides the resources for embryo development, a conflict can arise in the embryo between genes of maternal and paternal origin. The maternal genes in a given embryo will also be present in the female's future offspring; they are thus selected in order to spare resources for producing future embryos. On the contrary, there is no guarantee that the paternal genes will be present in this female's future offspring; they are thus selected in order to attract the maximum amount of resources, even at the expense of the mother's survival or future fertility. Conflict will occur only if the embryo can determine, as in mammals and higher-order plants, the amount of resources the mother allocates. (Oviparous mammals like the duck-billed platypus "fill" the egg before fertilisation so that paternal genes cannot influence resource allocation.)

Certain genes of the mouse display differential expression depending upon their origin, maternal or paternal. The gene for producing insulin-like growth factor II (IGF-II) which is a hormone that increases the resources allocated by the mother to the embryo is an example. During early development, the paternal genes leading to IGF-II production are expressed whereas the homologous maternal genes remain silent. On the other hand, the maternal, but not paternal, genes that repress IGF-II production are expressed. This is the reason why an embryo does not develop after the fusion of two ovules (to yield a progenote) or the injection of two spermatozoa into an enucleated ovule (to yield an androgenote). In the progenote, the genes from the ovules derive no resources from the surrogate mother; in the androgenote, the father's' genes are too costly. Only mixing the two genomes provides a balance between the "bulimia" of the paternal genes and the "anorexia" of the maternal genes.[5]

Do the above examples of conflict between maternal and paternal genes occur within an individual or between individuals? Even though the conflict takes place in the embryo, the protagonists are genes from the mother and father ... We have reached yet again the very limits of the notion of genome conflict.

## Transposons

The propagation of transposons illustrates another such limit. Transposons are DNA sequences about several hundred to several thousand bases long that code for at least one protein that enables them to replicate and invade a genome. The most widely studied transposon is element P, which, since the end of the 1940s, has invaded the genome of the fruit-fly (*Drosophila melanogaster*). It started out from an area known as the "Bermuda Triangle" and is now found in fruit-fly populations in America, Europe, and the Far-East.

Transposons are selfish. Their presence cannot be explained by any direct benefit to the host genome. On the contrary, they decrease reproduction by inducing partial or total sterility. Many mechanisms controlling their numbers and adverse effects are known and, paradoxically, most originate in the transposon itself! Can we therefore speak of conflict? For example, for a fruit-fly gene to be in conflict, it would have to repress the proliferation of P elements. Such repressor genes no doubt exist but have not yet been identified. It is thus not always clear from the outset whether a situation is really one of conflict or not. A selfish gene does not necessarily induce a conflict if it suffers as much as the host from the detrimental effect it induces.

Do P elements, or transposons in general, belong to a genome? In many organisms there are transposable elements that can move around or duplicate within a genome. (The sequence is copied and maybe excised. In some cases, duplication occurs via an intermediate RNA that then undergoes retrotranscription). These elements live up to their name of transposons and are rightfully considered to be autonomous. Some species possess inactivated transposons "frozen" into the genome which can be awakened for example by genetic engineering. Unfortunately, we know little about them. Transposons have sequence homologies with viruses. Making a clear distinction between retroviruses and retrotransposons, which both use an intermediate RNA, can be hard. Does one derive from the other or do they both derive from a common ancestor? Despite the possible relationship, we call viruses - which insert their genome into that of a host - "parasites" whereas we see transposons as elements belonging to the genome. Once more, the notion of a genome and thus of an individual is rather hard to define.

We have to revise the idea that the genome is a coherent assembly of elements, which, even if they are imperfect, contribute toward each other's success. The genome is more like a collection of pieces of information, each of which goes through a selection process which may be "natural" but which is not always geared toward the organism's adaptation. All of life's integration levels - and not just the organism level - are subject to selection forces. At some levels, the compartments are in conflict; at others, they play for the same evolutionary stakes and thus function in harmony. They collaborate often enough for organisms to adapt.

Overall, it is rather remarkable that a genome which is often made up of rather divergent entities manages to function and produce organisms with such subtle workings. This paradox is often studied in ecology. Colonies of living creatures are maintained despite the presence of parasites and predators. Just as, in ecology, we accept that interactions among species can prevent ecosystems from functioning optimally, we must suppose that genome conflicts might explain some of the imperfect ways in which individuals function and even some of life's other contradictions.

## 2. LET'S PLAY![6]

The reader will have understood by now that this chapter and the previous one are crucial to our thesis. We started off with a reductive, even reductionist, approach (freeing genetic information from the yoke of the nucleotide sequences in which it has been imprisoned by molecular biology) and have (re)introduced life's levels in a gradual, articulated, and ranked fashion, thus rediscovering "matter", organism, population, *etc*. Thanks to the ground we have covered, we now have a different outlook on biological reality. Less innocent or more pessimistic, perhaps, but, in any case, more intellectually "honest"... We are now ready to explore another of life's dimensions - games!

The mathematicians Johann Ludwig von Neumann and Oskar Morgenstern introduced game theory, as it is known today, in the mid-20th century in order to model the running of economic and military systems and predict decisions in a market economy. Decisions taken by one set of people influence actions taken by others, thus causing conflicts of interest. Is the best strategy the one giving the result that satisfies all players best? A moot point indeed! Game theory assumes that certain rules govern players' gains and losses as the game proceeds, the players know and accept the rules, and each player defines their own strategy at the outset and sticks to it throughout. The players base their strategy on the rules of the game and on what they know about the other players' strategies; they can even act in collusion with other players.

## **The chicken game or prisoner's dilemma** (see appendix MA6)

In the chicken game, two streams of cars heading in opposite directions have to tackle a bottleneck. What is the best strategy? Once one of the streams of cars has forced its way through, the drivers in the opposite stream cannot overturn the situation. Two equilibria are thus possible, each with a so-called stable strategy. If a driver deviates unilaterally from this strategy, he loses out.

In the prisoner's dilemma, two players are supposed to have committed a crime together. Each prisoner has the choice of either admitting to the crime or denying having taken part. He (or she) is, of course, unaware of his accomplice's decision. The following rules apply to both "prisoners":

- if both admit to the crime, their sentence is light,
- if both deny having taken part, they are both released,
- if one denies and the other admits, the one who admits will be released and rewarded, whereas the one who denies will receive a heavy sentence.

Let us imagine we are one of the prisoners. If we think our accomplice will deny, we have nothing to lose in admitting to the crime; we will even be rewarded. If we think our accomplice will admit, we have to admit also because, if we deny and our accomplice admits, we will receive a heavy sentence. In either case we have to admit to the crime because denial is too risky (we have assumed that we want to play safe!). Thus, the only sure and stable strategy is to admit. This strategy does not lead to the maximum gain either overall or for each player (because they have the same strategy at equilibrium); the gain is even negative. In such a context therefore, self-interest does not provide for an optimal overall outcome (as economists have long known, the "invisible hand" of Adam Smith is sometimes not very experienced). It is highly frustrating because all that is really needed is for the prisoners to agree to deny having taken part in the crime. Alas, in practice, it is always in our best interest to betray.[7]

## Games in evolutionary genetics

R. Lewontin was the first to apply game theory to evolutionary genetics but J. Maynard-Smith and G.R. Price were its chief proponents.[8] The basic idea was that cohabitation among congeners was just a *modus vivendi* between competing individuals. In models of evolutionary mechanisms, the players are the genes, that is the basic units of genetic information borne by the chromosomes. At each reproduction cycle, the genes are at start of play. They play to win the number of copies of themselves that will be present at each new generation, which we have called *fitness* or selective value.[9] Contemporary genes are those that have defined and implemented the best strategy over several thousand million years of evolution, and won. Just as the prisoner had to choose without knowing what his accomplice had decided, an individual's genotype is determined at conception, before entering play and before knowing the strategy adopted by the others. J. Maynard-Smith and G.R. Price added to these definitions the concept of Evolutionary Stable Strategy (ESS).[8,10] A strategy is evolutionarily stable if it establishes itself as the best for the individual. When the entire population has adopted this strategy, no deviant strategy (arising, for example, by mutation or from another population) can invade the population.

Under these circumstances, how can we account for rare strategies? How can we, once more, explain genetic polymorphism? One answer is to imagine some kind of "protection". If the frequency of an allele, initially close to zero, increases over generations rather than falls to zero, it is said to be protected.

## **A bourgeois butterfly** (see appendix MA6)

Game theory has been used to study how aggressive behaviour (hawk strategy), non aggressive behaviour (dove strategy) or territorial behaviour (bourgeois strategy) are selected. We can demonstrate the following theoretically. If the advantage of owning a resource (such as food or a female) is greater than, or even equal to, the risks of confrontation, the ESS is aggression. On the other hand, if the resource is worth less than the possible acquisition cost, then the ESS is bourgeois, that is aggressive only if one came in first.

Various species have adopted a bourgeois strategy. For example, several primates are bourgeois with regard to ownership of females. N.B. Davies has described bourgeois behaviour in a butterfly, *Pararge aegaria*, which lives in the undergrowth of forests.[11] For the males of the species, it is a considerable fitness advantage to take possession of a pool of sunlight on the forest floor because these pools attract virgin females. When a male butterfly arrives first at a light pool, it quickly drives away all other male butterflies that come close; these do not insist. If we set up a situation where two butterflies have claims on the same spotlight (by hiding the first butterfly when introducing the second, which therefore thinks it was there first), the confrontation lasts a very long time, each butterfly acting as if it were the owner of the patch.

## **ESS and sex-ratio**

We have already addressed the issue of sex-ratio and equality in male and female numbers. In a panmictic population of infinite size, the evolutionary

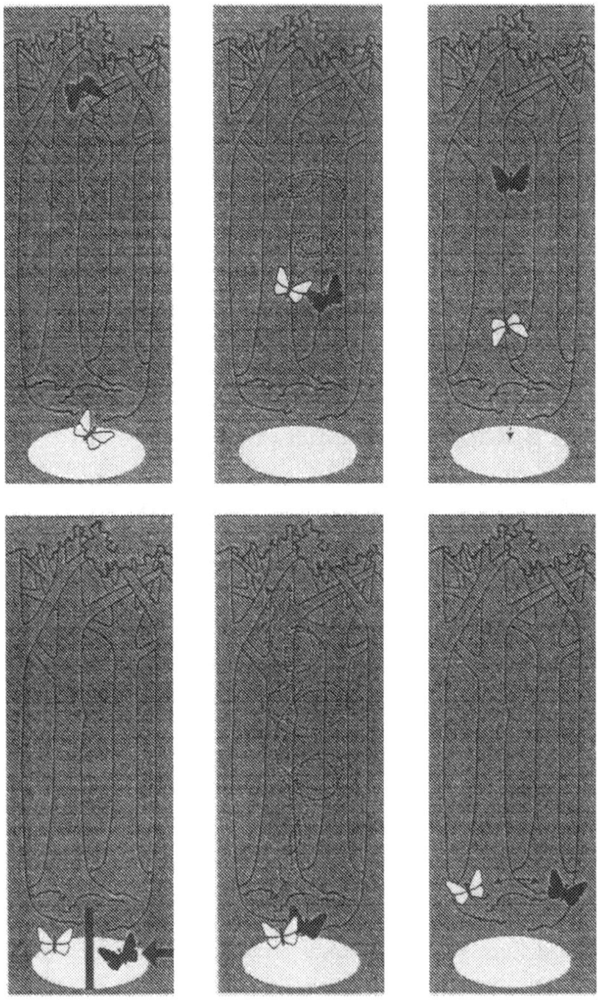

*Figure 5.3.* Example of bourgeois behaviour in the *Pararge aegaria* butterfly. In the top panels, a male swoops down from the tree tops towards a pool of light that is already occupied. The owner challenges him and, after a short spiral flight, drives him away. The intruder acknowledges his adversary's priority of claim. In the bottom panels, a butterfly is placed in the light pool unaware that the first butterfly is already there; both consider themselves to be owners of the spotlight. They then make a very long spiral flight which ends in the victory of both. The bourgeois does not try and obtain a resource that already belongs to another by force but fights to defend what he believes to be his own property. (Adapted from J. Maynard-Smith, The evolution of behaviour, Pour la Science, Nov. 1978; drawing by A. Brotmann).

stable strategy with regard to sex ratio is the production of as many males as females because this is the only strategy that provides for equal fitness of males and females. However, the fitness of the total population with such a strategy is below optimal; a population in which females outnumber males would have better fitness. The female sex is the one that is physiologically limiting with regard to the number of offspring. Breeders of domestic livestock are well aware that for maximum reproduction they must not keep all the males of a herd.

Populations in which related individuals mate can reach near-optimal fitness. If mating occurs within small units, the fraction of males to which the females need give birth decreases because of the inclusive fitness (provided by relatives). The smaller the number of founding females in the mating unit, the more likely that the female offspring of a given female will be fertilised by her brothers, who will pass on some of the mother's genes. The conditions are close to those of autogamy (see Chapter 4) where selection reduced "waste". The females are thus selectively trained to have fewer sons. If there is just one female, she will produce just the number of sons needed to fertilise all her daughters. This is called local mate competition.

## Evolution of seed dispersal rate

In some species, each individual can produce two types of offspring, those that do and those that do not leave the population or parental site. We can study how parents behave toward these two types of offspring.

Each plant of some species, especially of *Asteracae (Compositae)* such as thistles, produces seeds with or without pappa. The former are dispersed by the wind; the latter remain in the flower-head until it falls to the ground. The percentage of each seed type that a plant produces seems to be determined genetically but varies among plants since there is always some genetic variability in natural populations. We can use game theory to compute how selection determines the fraction of offspring that a plant or animal will disperse in order to colonise new habitats and the fraction that it will keep.[12]

Work by W.D. Hamilton and others has highlighted real conflicts between generations when the needs of parents and offspring do not converge. Often, it is in the parents' interest to send their offspring far away whereas it is in the offsprings' interest to stay close to their parents. The outcome depends on the determinism of the migration rate. In plants, the mother plant "decides" whereas, in animals, the decision is often taken by the young.

### Strategies and rules of the game

Game theory also provides a formal description of how certain traits that determine the reproduction rates of individuals evolve. We speak by analogy of the strategy of genes, of pre-established rules and decisions. The idea that genes may have a strategy is not novel; in 1957 Conrad H. Waddington entitled his book *The Strategy of the Genes*. The models in this chapter (sex ratio, dispersion, bourgeois character) are very robust with regard to genetic determinism of the traits involved. The conclusions are unchanged as long as the process is Mendelian (no superdominance and no segregation distortion of nuclear and cytoplasmic genes at meiosis). We can use these models to determine possible evolutionary stable strategies for a given set of internal genetic constraints (population composition) and physiological constraints (compensation curves between traits). The strategies may be pure (a single phenotype per genotype), mixed with probabilities attached (several phenotypes per genotype, each with a probability) or conditionally mixed (several phenotypes per genotype, adopted according to circumstance, as in the case of the bourgeois strategy).

There is not always a best strategy *per se* providing a genotype with maximum gain. The chosen strategy usually depends on the one adopted by the other players, as in the examples of frequency-dependent selection. It is thus defined within the framework of the population. The game of life is, first and foremost, a game of what is or what is not possible. The gene, like all genetic information, works out its strategy in a space subject to constraints.

## 3. THE SPLENDOUR AND WOES OF SOCIOBIOLOGY

To speak of hawk or dove strategy, or even of bourgeois strategy usually provokes no particular reaction among population geneticists despite the direct reference to classes of human society. To describe a gene as selfish is somewhat less anodyne and, to some people, rather worrying. Is there not a danger that when Richard Dawkins uses this expression he is endowing genetic structures or information with an independent will? This controversy reminds us of the time when well-educated people in British, and later Western, society were indignant when told that man descends from the ape. Only now, R. Dawkins and his colleagues are telling us that organisms are just machines for gene survival. Do evolutionists always have to come up with images that may shock the layman? Intentionally provocative statements should not blind us but we must admit that discoveries in biology have upset our traditional view of the world and our faith in man's position in this world. If science considers humans as just one among many other species that have descended from previous species (E. O. Wilson repeated Darwin's sacrilege in his behavioural studies), must we not conclude that the human species, like other species, cannot escape from the grips of evolution? Such a conclusion influences our understanding of men and human societies. We shall examine several consequences using sociobiology as an example.

## The cuckold and sociobiologists

The vaudeville scene we are about to describe would be of no interest at all - especially in a book on the evolution of biology - if the three traditional roles, that of the unfaithful wife, her lover and cuckold husband, were not held by birds with the poetic name of mountain bluebird (*Siala currucoides*). In 1976 the biologist David P. Barash carried out an experiment on two pairs of mountain bluebirds in Mount Rainer National Park, Washington.[13] When the resident male was out foraging, he placed a stuffed male at a metre from the nest and then watched how the bird reacted to this set-up "adultery" on its return. Barash repeated the experiment three times, once when the nest was being built, during incubation, and once the eggs had hatched. The bird's behaviour was intriguing. The male reacted violently to the intruder the first time but less aggressively once the eggs had been laid and hatched. Barash even noted that, during the first experiment, one of the males turned against the female and "*pulled an undetermined number of primary feathers out of his mate's wing. Two days later, this female was replaced by another female with whom the resident male successfully reared a brood.*"[13] The male's paternity, noted Barash, is only threatened during this first experiment when the "cuckold's" aggressive behaviour is most violent. Aggressive behaviour and the reaction to adultery thus seem related to the reproductive stakes. Barash sees this as adaptation from an evolutionist standpoint:

> *The results are consistent with the expectations of evolutionary theory. Thus aggression toward an intruding male (the model) would clearly be especially advantageous early in the breeding season, when territories and nests are normally defended ... The initial aggressive response to the mated female is also adaptive in that, given a situation suggesting a high probability of adultery (i.e. the presence of the model near the female) and assuming that replacement females are available, obtaining a new mate would enhance the fitness of males ... The decline in male-female aggressiveness during incubation and fledgling stages could be attributed to the impossibility of being cuckolded after the eggs have been laid ... The results are consistent with an evolutionary interpretation.*[13]

Barash introduced a new concept, anti-cuckoldry tactics, into studies of sexual behaviour.[14] It is the biological basis of a double standard. Males are sexually less discriminating, more aggressive, and more available than females. They put up with the unfaithfulness of their wives less than wives put up with the unfaithfulness of their husbands. In other times, this could have been the moral of a La Fontaine fable but D. Barash is no poet nor moralist but a scientist. How should we then interpret this conclusion? Is it a view of non-human living creatures that is rather (too) anthropomorphic or is it a description of life in general that includes humans and might therefore refer to human societies? In other words, does sociobiology - for that's what it is - offer a picture of the animal kingdom in human terms or a picture of human society using animals as examples? Much of the controversy surrounding sociobiology is underpinned by this alternative which some people see as an ambiguity or dilemma. We shall come back to this later. For the time being, we shall go to the roots of this controversy which has strongly marked the relationships between science and society during the last quarter of the 20th century.

## First social Darwinism, then sociobiology

The use of theoretical tools taken from biology to study human societies dates right back to the very genesis of Darwin's theory. Darwin was inspired by livestock breeding and selection methods but was also influenced by the work of Thomas R. Malthus (1766-1834) on human societies. From the very start Darwinism was linked to models of social thinking. It is therefore not surprising that, even before Darwin extended his theory to the human species in *The Descent of Man* (1871), several authors had already applied the Darwinian notions of a struggle for life and selection to individual and collective human behaviour. The best known defender of social Darwinism was without doubt Herbert Spencer (1820-1903), an English engineer and philosopher. (He designed an electromagnetic machine in 1841 and was interested in the production of metal deposits by electrolysis). Spencer formulated the first laws of social evolution at the beginning of the 1850s and invented the expression "survival of the fittest" which Darwin borrowed. According to the American historian Richard Hofstadter, social Darwinism encapsulated valid arguments for accumulating capital wealth or opposing the implementation of public aid programmes for the underprivileged. Obviously, such views induced reactions. Reformist Darwinism refers to the views of people for whom it is more natural to cooperate in life than to struggle for life.

The historian Linda Clark of Pennsylvania University notes that the expression "social Darwinism" acquired pejorative connotations at the end of the 1880s and during the 1890s.[15] However, a debate, with obvious political undertones, was ongoing before then, especially in France where the situation was more complex than elsewhere because of Lamarck's legacy. In *Les Colonies Animales* (1881) Edmond Perrier, a good example of the French scientists with bias for Lamarck, described the division of labour and unions as natural laws that were much more important than Darwin's struggle for life. However, the scientific reasons he gave in support of his view actually concealed very strong nationalist feelings inspired in large part by France's defeat in the war against Prussia. His work also echoed the concerns of many catholic republicans and conservatives who wanted to emphasise social cohesion rather than a class struggle in a country torn since 1789 by social crises. Some of the republican or socialist statements of the time nevertheless used Darwinian phraseology. A prominent defender of social Darwinism was Georges Vacher de Lapouge (1854-1936), a French lawyer and doctor, who suggested, in the name of anthroposociology, that the French motto "Liberty, Equality, Fraternity" be changed to "Determinism, Inequality, Selection". He developed theories on racial inequality together with Charles Richet, a French physiologist who was awarded the Nobel Prize for medicine in 1913. However, the French eugenist movement was not of pure Darwinian obedience. Perrier and other Lamarckians thought that improving the environment could benefit the "French race".

In 1935, Alexis Carrel, a French doctor and biologist, published *L'Homme, cet Inconnu*. This book earned him far more renown than did his 1912 Nobel Prize in medicine for work on surgical techniques (vessel suture and organ transplantation). In the footsteps of Americans and Germans, Carrel declared his allegiance to social Darwinism and negative eugenism, and recommended artificial selection in humans. He wrote:

> *Eugenism is necessary to perpetuate an elite. It is clear that a race must reproduce its best elements [...] No human being has the right to impose a life of misery on another human being and even less the right to procreate children who will be unhappy. In fact, eugenism demands the sacrifice of many individuals [...]. Sacrifice is an absolute social need. This idea must penetrate the mind of modern man.*[16]

A later passage in his book addresses euthanasia with appropriate gases in institutions.[17] Carrel was therefore part of a movement which fostered the development of scientific racism in France in the 1940s and which was upheld by the Vichy government and the Nazis who created the Institute of Anthroposociology.

## From entomology to sociobiology

This period of history, the 1940s, might have led biologists to be cautious when speaking about human beings but, of course, history does not remember those who were cautious but who, rightfully or wrongfully, trespassed upon forbidden ground. In the 1960s, the zoologist Desmond Morris did not shy away from observing humans with methods that took them for "naked apes". Edward O. Wilson, an entomologist from Harvard, was even more daring when he claimed that his knowledge of anthills helped him understand human societies. We shall trace Wilson's views and the ensuing debates using his major publications as milestones.

### *The Insect Societies* (1971)

Wilson published *The Insect Societies* after writing many papers on diverse aspects of ants (behaviour, chemical communication, polymorphisms, *etc*). The book established his position as an expert on insect societies, and its conclusion heralded what was to become a first extension of his work, a unified theory of sociobiology.

C.F. Hockett had introduced the term sociobiology in 1948 to describe a unified science. It supplanted rival terms such as "bio-sociology", "eco-biology" or even "bio-cultural theory of evolution". The aim of this new discipline was to explain how social behaviour patterns emerged and persisted, by postulating an evolutionary mechanism that we might call a local optimising process. This mechanism depended on the fitness conferred by a given social behaviour upon a population under specific constraints. William Hamilton's hypothesis, mentioned at the end of Chapter 4, was to play a crucial role.

Entomologists asked themselves how altruist behaviour can develop and persist in insect societies when it apparently brings no benefit to the altruistic insect and when the information determining this behaviour (genetic inheritance) is not transmitted directly. According to Hamilton's hypothesis, the adaptive value of an individual is measured not only from its personal reproductive success but also from the benefit it procures from the reproduction of a third party. This applies as long as it has a substantial part of its genome in common with the third party, that is, if it is a close relative.[18] The hypothesis explains fundamental behaviour patterns such as altruism but also less "generous" behaviour such as aggression, rivalry between the sexes (as observed in mountain bluebirds!) or the behaviour of social parasites. W. Hamilton called inclusive fitness the sum of the altruist's selective fitness and the indirect genetic benefit to its genes due to altruism toward a parent. This is the basis of the kin selection mentioned in the previous chapter.

E. Wilson studied animal strategies chiefly in terms of the costs and benefits in adaptive value and used this principle to calculate the optimal number of insects in the teams that are assigned various duties within an ant colony.[19] This was just a first step towards a synthesis. At the end of the book, he wrote a chapter on insect societies:

> *These two groups of animals [vertebrates and insects] have evolved social behaviors that are similar in degree of complexity and convergent in many important details. This fact conveys a special promise that sociobiology can eventually be derived from the first principles of population and behavioral biology and developed into a single, mature science.*[20]

We should thus be able to extend the systematic study of the biological basis of social behaviour, as developed by Wilson for ant populations, to all living beings including Vertebrates (Wilson mentions rhesus monkeys). So far, no objection ...

### *Sociobiology: The New Synthesis* (1975)[21]

This work is a well-illustrated and documented encyclopaedia on animal societies from the Invertebrates to humans. Wilson affirmed that systematic study of the biological basis of social behaviour patterns would give rise to a "new synthesis". A basis for the synthesis was W.D. Hamilton's view: the aim of altruist behaviour in animal species is not to ensure the good of the species nor of the altruist but of a relative who indirectly propagates the altruist's genes. E. Wilson added two models from Robert Trivers (1971) on reciprocal altruism and parental investment.

Reciprocal altruism is a kind of social contract between individuals who help each other without trying to take advantage of each other's kindness. Wilson gives the following definition:

> *The trading of altruistic acts by individuals at different times. For example, one person saves a drowning person in exchange for the promise (or at least the expectation) that his altruistic act will be repaid if the circumstances are reversed at some time in the future.*[22]

The hypothesis led to several objections. Apart from the fact that, at sea, it is dangerous to put ones fate in the hands of someone who cannot swim (but why should the rescued person not save his benefactor from fire some other time?), there is the problem of the "first step". Reciprocal altruists have to perform several charitable acts at a pure loss become deriving any benefit and, especially, any selective advantage. On the other hand, as the theory's opponents have stressed, a selfish person will prosper at an incredible rate in a society of reciprocal altruists. E. Wilson and other sociobiologists admitted that there were prerequisites to reciprocal altruism such as a rather highly evolved society or, at least, relationships that are close enough for someone who takes advantage to be "punished". Notwithstanding, the hypothesis might explain the behaviour of baboons in Tanzania or of vampire bats.[23]

The notion of parental investment has also raised objections. It tries to account for rather strange happenings such as the female murdering the male during or after mating. This is what the praying mantis and some spiders do:

> *Murder of the male is no accident. It is not due to a failure in the inhibitory system of the partner's predatory behaviour. It is due to the fact that these are carnivorous species and that males are part of a female's diet. Cannibalism is adaptive for the animals of both sex since the extra food acquired by the female will enable her to bear more offspring. It is not the male's intention to sacrifice himself. The males that managed to escape have been eliminated by natural selection. By providing less nourishment to the female who bears their heirs, they invest less in their reproduction.*[24]

In the hypothesis of parental investment, each parent seeks to gain the maximum return for its investment in offspring. Because the female gamete is larger and the female nourishes the embryo, maternal investment is far greater than male investment. Females have to tend and feed the young just because investment by the two sexes is so disproportionate; even if they have to make up for it by eating the male! This was the idea - introduced as a matter of course in the chapter on sociobiology - that sparked things off. After 26 chapters devoted to animals, E. Wilson started speculating about the genetic determinism of human behaviour which he took to be universal. Why not use the theory of parental investment to explain the status of women in most human societies? And why not see social behaviour as an expression of gene ethics, of a strategy to enhance reproduction, whether we are dealing with religion, culture, the existence of social classes, conflicts, *etc.*? Should we not invent a new discipline called genetic anthropology to explain the specific, varied features of human behaviour? In reply to a critic, E. Wilson wrote in the *New York Times Magazine* (October 12, 1975):

> *There is no doubt that the patterns of human social behavior, including altruistic behavior, are under genetic control, in the sense that they represent a restricted subset of possible patterns that are very different from the patterns of termites, chimpanzees and other animal species.*[25]

To proclaim such ideas is to "*put man in the accusation box*",[26] but also to go along with Jean Rostand who admitted that he was "*more inclined to find the beast in man than to seek the man in the beast*".[27] Whether the last chapter of *Sociobiology* refers to the first or second proposition is difficult to say. Is it a picture of animal society in anthropological terms or is it an analysis of human society based on data from animal populations? E. O. Wilson chose to elaborate on a human sociobiology.

### *On Human Nature* (1978)[28]

*On Human Nature* describes the human species in rather unflattering terms. Genetically, it is bellicose and prone to blind faith in a divinity or in its leaders. Relationships between men and women are on a par with those of D. Barash's mountain bluebirds. Is there not a tendency for man to be polygamous and jealous, and for a woman to be faithful and to care for her children, in line with the theory of parental investment? Sociobiology, if generally applicable, should help human societies identify the genetic basis for their behaviour patterns, whether these are common or rare, whether they concern sexuality or culture. If these patterns are the products of an evolutionary process, why not try changing and amending them, within certain limits and in certain directions, of course? To those who took him for a dangerous conservative, E. Wilson retorted that he championed a science that welcomed all hypotheses and did not defend any ideological social status quo.

Wilson stuck to his guns in *On Human Nature* but his adversaries did not quit the battlefield. The most constructive criticisms probably came from biologists, in particular from his colleagues at Harvard University. Richard Lewontin and Stephen Jay Gould, a palaeontologist, were the leaders of an opposition fostered by a very small Marxist group, *Science for the People*. They countered Wilson's genetic determinism with the idea of biological potentiality. For example, S. J Gould wrote:

> *Violence, sexism, and general nastiness are biological since they represent one subset of a possible range of behaviors. But peacefulness, equality, and kindness are just as biological - and we may see their influence increase if we can create social structures that permit them to flourish.*[29]

This opinion could be called "environmentalist" but is nevertheless strictly biological. According to S.J. Gould, the human brain offers a wide range of possible behaviour patterns which are *a priori* equivalent and which form a kind of biological potentiality. He agrees that the brain is a purely biological product but rejects the straightjacket of determinism. Wilson will not.

***Genes, Mind and Culture*** (1981)[30]

In *Genes, Mind and Culture*, Charles Lumden and E. Wilson introduced a modified version of human sociobiology. They adopted a new working hypothesis, co-evolution. Genetic information, which is subject to natural selection, determines the mechanisms or epigenetic rules that will affect selection and how cultural traits (which they call *culture genes*) are filtered. All practices, such as making a harpoon, wearing a long or short skirt, having brother-sister incestuous relationships or believing in God, came into being and were maintained by the processes of invention, accident and diffusion. The rate at which a cultural trait is adopted (each trait has a fitness) and the frequency of genes programming the selection of certain traits thus change at each generation.

E. Wilson and C. Lumsden tried to avoid the trap of the culture/gene or even acquired/innate antitheses. They did not consider the determinism of these *culture genes* as entirely genetic but emphasised that genes keep culture "on a leash". Michel Veuille notes that neither the sophisticated mathematics of this book nor its use of statistics can hide the authors' oversimplification in considering both the genome and culture as mere sums of distinct elements.

***Promethean Fire, Reflections on the Origin of the Mind*** (1983)[31]

In this book, E. Wilson and C. Lumsden omit the mathematical formulae thus enhancing the clarity of their message. The image of a man dancing around the DNA double strand at different stages of his life is a good description of their opinion; evolutionary progress and genetic determinism are at the centre.

## Will (socio)biologists seize power?

S.J. Gould and R.C. Lewontin were not alone in criticising the opinions of E. Wilson and the sociobiologists. The criticisms by experts in the social sciences were, however, of another sort. Did they not have the right to attack researchers who, although not competent in their field, had the nerve to teach them lessons? Such reactions were understandable, says the French sociobiologist Pierre Jaisson, since:

> *some of the clumsy statements and insufficiently detailed claims made by the first sociobiologists had complicated the situation.*[32]

Nevertheless, in line with the principle that "*no science can be judged without evaluating its ability to furnish concepts that might enhance our knowledge*",[33] the controversy gave way to debate and dialogue, especially in the United States.[34] For instance, in *The Evolution of Altruism and the Ordering of Love,*[35] Steven J. Pope admits that Christian moralist theologians had until then ignored or rejected outright the opportunities offered by sociobiology studies. His plan was to use these studies to analyse and lay the foundations for a pillar of Christian morals, the Gospels' command to love thy neighbour.

Sociobiology had a quite different standing in France which P. Jaisson summarises in a single sentence. Quite simply, sociobiology had been forbidden as a scientific discipline because of its "*amalgamating logic that reduced the meaning of a theory to its undesirable effects which presumably could not be avoided.*"[36] Sociobiology had only existed under the auspices of the "Nouvelle Droite" (New Right) which transformed it into "sociobiologism". In 1981 Pierre Thuillier wrote "*Will biologists seize power?*" [37] but by biologists he really meant sociobiologists. These reactions had two consequences. Competent scientists quit the field and the French anti-Darwinian flame that had waned since the mid-20th century was rekindled. Whatever the substance and outcome of these debates, their merit was to make sociobiologists cautious. The Statutes of the European Sociobiological Society clearly state that:

> *The Society will, in accordance to its statutes, refrain from using or abusing such studies for political purposes.*

Caution, however, should not only be exercised in deontology and ethics. The sequel to the mountain bluebird story underscores the need for caution in science itself.

## **The cuckold and sociobiologists** (contd)

D. Barash's claims may have been daring, correct or excessive, but his experiments had shortcomings. For example, the male bird might have had ever weaker reactions when repeatedly confronted with the stuffed bird simply because he had become accustomed to his rival. Additional experiments were needed to confirm or refute the first results. In 1978, Morton, Geitgey and MacGrawth examined the behaviour of the Western Bluebird, a species close to the Mountain bluebird, but, to quote Michel Veuille, found no "Blue-beard". On the contrary, the females turned out to be highly aggressive in counter-experiments with female decoys. Thus, D. Barash's theory was probably not as widely applicable as he had thought. In 1980, Power and Doner repeated the experiments with the Mountain bluebird but increased the number of couples to 41, using either stuffed birds or birds in cages (these proved to be more reliable). Yet again, the results were unexpected. The females attacked the female intruder more often than the males attacked the male intruder! The researchers interpreted this as a defence of the nest, a behaviour common to both sexes which does not wane with time. Each partner preferred to attack its own sex probably not because of anti-cuckoldry behaviour against a rival but rather because it wished to avoid upsetting individuals of the opposite sex which were potential mating partners. Anti-cuckoldry tactics only prevailed when the nest was being built; at this time the male did not budge from the female's side. Does Power and Doner's work contradict the sociobiologists' theses? Perhaps not, because it might offer "*another form of sociobiology, with a more rigorous method and less outrageous extrapolations*".[38] The cuckold in this story might be the sociobiologist who was a little too quick off the mark to believe that he had a good illustration of evolutionary adaptation!

Our overview of sociobiology is not meant to be comprehensive. The reader can refer to the works we have cited and in particular to the work of the French sociobiologist P. Jaisson who provides a critical and well documented introduction to these studies as well as keys to understanding the ideological and sociological dimensions of sociobiology. To conclude this chapter, we shall expand upon a few of them.

## Weighty words, shocking snapshots

By snapshots we mean sociobiology's present-day image. This image varies a great deal depending on country and on whether biologists took any notice of sociobiology. The Anglo-Saxon world gradually integrated E. Wilson and his colleagues' explorations into their thinking whereas France reacted to the sociobiologists' theses by brutally rejecting their work. We should not lay too much emphasis on the exceptional French reaction. Its origin lies in three factors with different impacts: the history of social Darwinism and possibly of its political links with Nazism in France; the clash between Lamarckism and Darwinism, and the changing relations between "Science" (or scientific statements) and society. We have already mentioned the first two factors; the third deserves a lengthy digression. Let's see how it fits in with the three revolutions due to modern science outlined by Pierre Teilhard de Chardin:

> *Ever since Galileo (according to the remark by Freud), in the eyes of Science, Man had not stopped losing, one by one, the privileges that, until then, had made him seem unique in the World. First, in the eyes of*

*astronomy because, like the Earth and with the Earth, Man was lost in the giant anonymity of the stellar masses; then in the eyes of biology because, like any other animal, Man was lost in a multitude of sister species; finally, in the eyes of psychology because in Man's heart there gaped an abyss of Unconsciousness. In three successive steps and in four centuries, Man, you hear, seemed to have dissolved definitively into the common mould.*[39]

Is there a scientific discourse that can claim to be perfectly innocent in this regard? Even when emphasising science's benefits to Mankind, it will always reflect the gradual loss in the privileges that Man believes he once had. The words and images scientists choose to use matter a lot in such a context. No longer can we describe biological processes with words reserved to describe humans (in line with the first proposition of the question "does sociobiology offer a picture of animal society in human terms or a picture of human society using animals as an example?") and yet continue to claim that we adhere to the ancestral separation between human and non-human. Is not the modern scientific outlook more interested in the differences than similarities between Man and animal?

Those who defend a "traditional" human identity thus call into question not only the claims made by D. Barash but also notions such as "genetic information", expressions such as "selfish gene" or "selfish DNA", and the idea expressed in the shock-phrase "individuals are artifices that genes have invented for the purposes of their own reproduction". R. Dawkins specifies in *The Selfish Gene* that *"We, alone on earth, can rebel against the tyranny of the selfish replicators".*[40] Maybe so. However, sociobiology and the theory of selfish replicators are indicators of the breach to which contemporary biology has led us along the "revolutionary" road taken by Copernicus and Galileo.

By placing sociobiology's discourse in a broader epistemological perspective, we do not, of course, wish to excuse or forget the verbal excesses and unfortunate, unsubtle statements that marked sociobiology's first steps and which have not completely disappeared. In a way they illustrate the social and ideological conditions from which scientific discourse cannot abstract itself totally. It would be equally foolhardy to hope to avoid misconceptions. P. Jaisson has recorded the main ones:[41] "Sociobiology is predetermination of behaviour by genes", "Sociobiology is sexist", "Sociobiology is racist" "Wilson is totally isolated and discredited". He provides a critique of each of these statements, but is it possible to avoid what Pierre-André Taguieff calls "*amalgamating logic*"?

The second proposition "does sociobiology seek to describe human society using animals" is an ongoing question that still needs an answer.

## What science for Mankind?

The last chapter of E. Wilson's *Sociobiology* (as well as that of *The Ant and the Sociobiologist* entitled "Man and the sociobiologist") should not be overlooked. It claims to offer keys to the understanding of human behaviour based on work on animal societies. The Parisian biologist Jean-Michel Goux has noted that it is no longer a question:

> *of combining trains of thought that until then had been separate, but of the approach taken by some biologists - not necessarily geneticists - who tend to disqualify sociology as a science, as knowledge, on the basis of a theory of individual behaviour. These biologists simply tend to turn the field of reality made up by human societies into an annex of biology and refuse to endow these societies with any intrinsic specificity of their own compared to animal species.*[42]

Goux underlines the convergence between this view and social Darwinism. Sociobiological views of our understanding of Man have raised, beside rather unfortunate historical echoes, questions on three themes: the competence of sociology, the rationale of the reductionist approach, and finally the opposition between nature and culture.

The issue of sociology's competence can be viewed as yet another one of the battles waged between various fields of human knowledge and experience that purport to understand Man. These fields include *de facto* all aspects of anthropology: mythological, philosophical or theological anthropology, political or economic anthropology, *etc*. When a branch of biology takes social sciences, in particular sociology, to task, this obviously generates friction and reactions. All systems of thought or knowledge tend toward conformity and react strongly to what might threaten their fields of competence and influence. In short, some of the reactions provoked by sociobiology are "fair play"!

The Canadian philosopher Michael Ruse takes this view when describing the conflict between sociobiology and the social sciences. He places the conflict issue in the broader context of a passage from one theoretical system to another and is particularly interested in the instance when an old system becomes just an example or a specific case of a new system. The substitution occurs by a process of reduction. The term reduction may be unwelcome especially when applied to the so-called "social" sciences. Can they be reduced to a specific case in biology or, worse, in genetics? However, as noted by Ruse, reduction can be an opportunity for a truly creative experience that need not involve the "disappearance" of those who have worked on the non-reduced theory.[43] The three revolutions invoked by

Teilhard de Chardin (that of Galileo, Darwin and Freud) belong to this type of process. They were both hurdles and inspirations for human thought.

M. Ruse then notes how the anthropological study of societies that have remained in intimate contact with nature is sometimes close to sociobiology. The idea is surprising but noteworthy. Of course, we do not wish to compare peoples in the South Sea Islands with anthills. The issue is the link between nature and culture, and the role of determinism.[44] The South Sea Islanders' relationship with their environment belongs to the realm of norms rather than aesthetics. Nature imposes its law and morals to all living creatures including humans. Individual and collective behaviour is thus strongly conditioned by what are seen to be natural laws. Does this mean that there is quasi-total natural determinism? Probably not, because interpreting nature and its meaning is a kind of culture. It is just that the distinction between nature and culture is more tenuous in the South Sea Islands than in Western society. Western society claims to have driven off natural laws in order to impose human laws, culture ... and fears that they will return disguised as sociobiology!

This fear is not only expressed in the rejection of the "good savage" myth. It also concerns determinism. Is not accepting the sociobiological approach recognising that genetics can determine our persons and societies? E. Wilson defends genuine genetic determinism but his opinion is far from being shared by all sociobiologists. As an answer to the accepted idea that sociobiology tells us that behaviour is predetermined by genes, P. Jaisson writes:

> *Most sociobiologists consider that animal and human behaviour patterns are made possible by their genetic potential. For them, genes do not predetermine but predispose to the development of behaviour patterns that will be put into place during epigenetic development under the influence of the individual's experiences. Animal sociobiology has made many discoveries on acquisition through experience especially as regards social or sexual partner recognition.*[45]

Behaviour is a reality that relates to the entire organism under defined environmental conditions. Maybe we should speak of potentiality rather than determinism. André Langaney defends the totipotent-nullity of the human newborn:

> *Unlike newborn animals, little Man is highly undetermined at birth. As he is not pre-adapted, he can adapt to nearly everything because he has a memory, and can integrate and re-use the information he has received. No other species can do this.*[46]

The notorious issue of the relative importance of heredity and environment, nature and culture, innate and acquired (these pairs are probably not equivalent) thus loses intensity in the fields of both social concourse and intelligence. M. Ruse suggests a compromise between biology and culture:

> *Human behaviour can thus be seen as biologically adaptive, which is what the sociobiologists want, but crucially causally influenced by learning, which is what the culturalists want.*[47]

Genes and environment interact in an extremely complex fashion and currently we can neither define nor analyse the interaction. One thing is certain, people are not molecules and there is neither fatality nor biological determinism in a strict sense. Neither is there any hope that the individual can free himself from life's constraints.[48] Even humans cannot escape from the difficulty of defining life in their efforts to understand themselves. This is the opposite route to the one taken by Schrödinger who after asking *What is life?* went on to ask what is "I".[49]

There is probably not one science of Mankind. Whether sociobiologists agree with this statement or not, we are clearly reminded of it by sociobiology's outrageous claims. More so than earlier currents of thought and theories, sociobiology invites research workers and thinkers, irrespective of their ideology, to set aside the opposition between nature and culture when explaining human abilities and behaviour. The weakening of inter-disciplinary boundaries may be the outcome of sociobiology's excesses but it is nevertheless chiefly due to current advances in genetics. The compromise is not an easy one, at least in absolute terms. In practice, maybe we should accept, with no negative preconceptions, the research work and results that refer to the one dimension rather than the other. This is not morally obvious.

## Morality here, there, and everywhere

Morality (or ethics, we shall make no distinction between them) is in fashion. At the turn of this century, have we decided to scare ourselves not with nuclear arms but with genes, transgenes, culture genes and other avatars of genetic information? This is not a criticism of Western society's scruples about the risks attached to either old (combustion of coal and petrol) or new (nuclear energy, biotechnology) technologies. Nor do we want to spread unease. But can we look at simple indicators such as those for world population, air pollution, or water quality and availability without being anxious? Our era has witnessed the birth and growing dominance of a new principle, the precautionary principle, which clearly rests on fear.[50]

Our society's relationship to ethics is still suffering from a double shock from World War II. Two spectres - death camps and the atomic bombs released over Japan - haunt our minds and explain our moral dread of sociobiology and of genetics. Each time we evoke these spectres, we have the fear of seeing them rise again under a different guise, endowed with an even greater destructive power. We thus forestall the risks inherent in our projects and activities, and stamp out, at inception, anything that might look like an eugenist statement or endeavour.

That ethics should lie at the heart of our society is no surprise. M. Ruse affirms that culture is not at the origin of morals but that, on the contrary, it somehow presupposes morals. If a human society does not lay down a minimum number of rules for cohabitation, there can be no culture. Does this imply that morality is the product of nature only (in the genetic meaning of the term) and, as such, subject to natural selection? Unless, in line with the reversive effect proposed by the philosopher Patrick Tort,[51] natural selection has ended up by selecting civilisation and its morality. But civilisation and morality are opposed to natural selection and to its avatars (as commonly understood!) which, according to its opponents, are eugenist practices and social Darwinisms.

As we have tried to show, the statements on human thinking inspired by sociobiology and other fields of biology must always be considered in relative terms. Human society cannot do without morality. But does it need overzealous defenders of morality? E. Wilson preferred to return to his ants and now restricts himself to the study of the extinction of species and the evolution of biodiversity.[52]

## NOTES

1. Chromosomes, the genomes of mitochondria, chloroplasts and endosymbiotic bacteria, as well as transposons and other repeat sequences within chromosomes, which are more or less independent, as we shall see.
2. Before meiosis, the genes that form haplotype t induce the production of harmful substances that will turn up in the cytoplasm of all gametes. Only the gametes that carry haplotype t can protect themselves against these substances.
3. Both in the wild and in laboratory "population cages".
4. Östergren, G., 1945, Parasitic nature of extra fragment chromosomes, *Botaniska Notiser* 2: 157-163 (p. 163).
5. See section entitled "Ad hoc developmental constraints" in Chapter 6.
6. This section is drawn from Gouyon, P.H. and Olivieri, I., 1989, Génétique évolutive et théorie des jeux, *Société Française de Biométrie* 8: 18-38.
7. This game illustrates the difference between a stable strategy (or "Nash equilibrium", admitting) and optimal strategy (or "Paretto optimum", denying).
8. Maynard-Smith, J., 1961, Evolution and the Theory of Games, *J. Theor. Biol.* 1: 382-403.
9. See the subsection on selection in Chapter 3.
10. Maynard-Smith, J., and Price, G.R., 1973, The logic of animal conflict, *Nature* 246: 15-18.

11. Davies, N.B., 1978, Territorial defence in the speckled wood butterfly (*Pararge aegaria*): the resident always wins, *Anim. Behav.* 89: 83-129.
12. Hamilton, W., and May, R., 1977, Dispersal in stable habitat, *Nature* 269: 578- 581.
13. Barash, D.P., 1976, Male response to apparent female adultery in the mountain-bluebird: An evolutionary interpretation, *Am. Nat.* 110: 1097-1101.
14. D. Barash supported this interpretation with observations on the mallard duck. When a male mallard duck discovers that his female has apparently been "raped", he quickly "rapes" her in turn in order to reduce the chances that her eggs be fertilised by the sperm of another.
15. Clark, L., 1984, *Social Darwinism in France*, University of Alabama Press.
16. Carrel, A., 1935, *L'Homme, Cet Inconnu*, Plon, Paris, pp. 287-289.
17. "*As for the others, that is those who have killed, committed armed robberies, abducted children, made destitute the poor, or taken advantage of the public's faith, they could be got rid of, quite humanely and economically, in an institution that performs euthanasia with appropriate gases. Moreover, would not the same treatment be appropriate for madmen who have committed crimes?* " (Carrel, A., *op. cit.*, p. 306)
18. See Jaisson, P., 1993, *La Fourmi et le Sociobiologiste*, Odile Jacob, Paris, p. 53.
19. In particular, he developed the concepts of r-strategy and K-strategy that have since been used to describe the explosive or moderate growth of animal populations.
20. Wilson, E.O., 1971, *The Insect Societies* (Chapter 22: The prospect for a unified biology), The Belknap Press of Harvard University, Cambridge, Mass, p. 460.
21. Wilson, E.O., 1975, *Sociobiology: The New Synthesis*, The Belknap Press of Harvard University Press, Cambridge Mass.
22. Wilson, E.O., *ibid.*, p. 593.
23. See Jaisson, P., *ibid.*, p. 122.
24. Veuille, M., 1986, *La Sociobiologie*, Series: Que Sais-je? N° 2284, Presses Universitaires de France, Paris, p. 51.
25. Quoted by Gould, S.J., 1991 (originally published in 1978), *Ever since Darwin, Reflections in Natural History*, Penguin Books, p. 252.
26. Grassé, P.P., 1980, *L'Homme en Accusation: de la Biologie à la Politique*, Albin Michel, Paris.
27. Rostand, J., 1978, *Pensées d'un Biologiste*, Stock, Paris, p. 236. Elsewhere, Rostand described man as a denatured ape, soluble in nature, or as a miracle of no interest.
28. Wilson, E.O., 1978, *On Human Nature*, Harvard University Press, Cambridge Mass.
29. Gould, S.J., *ibid.*, p. 257.
30. Wilson, E.O., and Lumsden, Ch.J., 1981, *Genes, Mind and Culture: The Coevolutionary Process*, Harvard University Press, Cambridge Mass.
31. Wilson, E.O., and Lumsden, C.J., 1983, *Promethean Fire, Reflections on the Origin of the Mind*, Harvard University Press, Cambridge, Mass.
32. Jaisson, P., *ibid.*, p.12.
33. Jaisson, P., *ibid.*, p. 291.
34. P. Jaisson has described the climax of the crisis, the "cold shower" that E.O. Wilson received during a symposium in Washington in February 1978.
35. Pope, S.J., 1994, *The Evolution of Altruism and the Ordering of Love*, Series: Moral Traditions and Moral Arguments, Georgetown University Press, Washington.
36. Taguieff, P.A., 1990, *La Force du Préjugé*, Gallimard, Paris, p. 68.
37. Thuillier, P., 1981, *Les Biologistes Vont-ils Prendre le Pouvoir? La Sociobiologie en Question*. Series: Le Contexte et l'Enjeu, Complexe, Paris.
38. Veuille, M., *ibid.*, p. 881.
39. Teilhard de Chardin, P., 1956, *L'Apparition de l'Homme*, Seuil, Paris, p. 367. Teilhard de Chardin bases himself directly on what Sigmund Freud said.
40. Dawkins, R., 1976, *The Selfish Gene*, Oxford University Press (new edition 1989, p. 201).
41. Jaisson, P., *ibid.*, p. 17.

42. Goux, J.M., 1985, Génétique et sociologie, in: *Misère de la Sociobiologie* (P. Tort, ed.), Presses Universitaires de France, Paris, p. 31.
43. Ruse, M., 1982, *Sociobiology: Sense or Nonsense?* Reidel, Dordrecht, p. 168.
44. Lenoble, R., 1969, *Esquisse d'une Histoire de l'Idée de Nature*, Albin Michel, Paris; Bourg, D. (ed.), 1993, *Les Sentiments de la Nature*, Series: Essais, La Découverte, Paris.
45. Jaisson, P., *ibid.*, p. 17.
46. Langaney, A., 1987, *Le Sexe et l'Innovation*, Series: Points Science, Seuil, Paris, p.105.
47. Ruse, M., *Sociobiology: Sense or Nonsense?* p. 160.
48. See Chapter 7.
49. See Section 3 "What is life?" in Chapter 4.
50. Jonas, H., 1979, *Das Prinzip Verantwortung*, Insel Verlag, Frankfurt-am-Main. The German philosopher Hans Jonas bases his ethics on responsibility toward future generations. Moral knowledge is rooted in knowledge of evil, passed on and made visible by fear.
51. Tort, P., *La Pensée Hiérarchique et l'Evolution* (1983), Aubier, and *Misère de la Sociobiologie* (1985), Presses Universitaires de France, Paris. Patrick Tort has critically reviewed statements on sociobiology and tried to show how far removed they are from Darwin's statements.
52. Wilson, E.O., 1988, *Biodiversity*, National Academy Press, Washington; Wilson, E.O., 1992, *The Diversity of Life,* Harvard University Press, Cambridge, Mass.

# Chapter 6

# EROS AND THANATOS REVISITED

We started this book by asking "What is life?" but immediately added that the ultimate reply is beyond our reach. However, this did not stop us from applying the tools of contemporary evolutionary genetics to an examination if not of life, at least of living things. In the previous chapter, we examined the relations among genomic elements, members of a population, and species of an ecological community. We underscored the tug-of-war between conflict and cooperation before inviting the reader to reflect on whether, and to what extent, these ecological results could be applied to human societies and persons. In this chapter, we shall also address relations among living organisms but from the angle of life's "drama"; wc shall look at sex and death.

In pre-Freudian times, Eros and Thanatos were a virtually inseparable couple. They were paired up for many reasons that were either mythological, symbolical or practical. Man's experience of love can be related to his experience of death; is not orgasm a "little death"? However, we must go beyond this anthropomorphic comparison and investigate the interdependence of these events, looking at what is actually passed on by sex and what is perpetuated in spite of death, in other words genetic information. Inspired, yet again, by Richard Dawkins, we shall update the chicken and egg question to conclude: Human organisms are the survival machines for genes. We need have no fear that this "reductionist" view will destroy the dramatic or poetic intensity of either death or procreation. Despite its simple wording, R. Dawkins' paradigm in no way reduces the diversity that characterises the living. Even from a gene's point of view, there can be many reproductive strategies and many options for avatar longevity.

# 1. DEATH AT OUR HEELS?

Is death an event of no importance, a simple consequence of time and chance, that has ultimately little to do with biological reality? Some people, like for instance MacFarlane Burnet, have claimed that there may be no genetic process that might delimit ageing in small vertebrates.[1] Others like Jean Hamburger have doubts about this:

> *The idea that death might be an "accidental" event in many animal species other than mammals seems to me more novel than certain. An event from which an animal never, absolutely never, escapes, often at a given time of his life, can hardly be considered accidental.*[2]

He adds that "*I find the idea that life's main secret is error appealing, even though it is paradoxical at first sight.*" The idea converges with that expressed by Jacques Monod in "*Le Hasard et la Necessité" ("Chance and Necessity").* If we pursued this confrontation between Burnet and Hamburger further, it would take us along a by-road to the questions we have already raised on the acquired and innate, and on the relative contributions of genetic inheritance and environment to the fate of an organism's body and mind. In this chapter, we shall choose the genetics route that is Hamburger and not Burnet. Death is not at the heels of organisms since it is already inscribed in them. In other words, the living are born to die. As Jean Didier Vincent put it:

> *Ageing is a hereditary disease that we carry in our chromosomes.*[3]

We shall thus address ageing, or more precisely senescence, which concerns both the individual as a whole and certain organs and cells in particular.

## Semelparity and iteroparity

In nearly all known sexual species, the life cycle begins with fertilisation and birth and moves on to the development of an adult individual which can reproduce its genes. Usually, once the age of first reproduction is passed, vigour and survival capacity diminish. This phenomenon is termed senescence. There is an extreme form of senescence called semelparity. A species is semelparous when the individuals die immediately after reproducing. This happens in many plants such as annuals (e.g., corn, maize or poppies), bi-annuals (e.g., beetroot) but also in some perennials which, having developed over several years, flower, produce seed, and then die. It also occurs in many insects. The male praying mantis, observed by Jean-

Henri Fabre (1823-1915), is a well-known example. The cicada spends several years as an underground larva before emerging, reproducing and dying. Butterflies and moths also die immediately after reproducing.

**The courtship of the praying mantis**

*We are nearing the end of August. The male, a lover of slight build, thinks the time has come. He glances several times meaningfully toward his powerful companion, turns his head towards her, bends his neck, and puffs up his breast. His roguish little pointed face has a look that is nearly passionate. Without moving, he observes the object of his desire at length. Although she does not budge, as if she were indifferent, her lover has noted some secret sign of acquiescence which I cannot fathom. He moves in closer and frequently spreads his wings which quiver convulsively, a sign that he is popping the question. He then thrusts himself upon her, his puny body on her portly back, holds on as best he can, and steadies himself. Foreplay usually lasts a long time. Finally, they mate lengthily sometimes for five to six hours.*

*There is nothing to note whilst the two partners keep still. At last, they separate but come together again in a more intimate fashion. She likes the poor little thing not only because he brings her ovaries to life but because he is a tasty morsel. Thus, on the very same day or the following day at the latest, she grabs him and, as custom will have it, gnaws at the nape of his neck and then methodically eats him up, bit by bit, leaving only the wings. This is no case of jealousy in a harem but downright depraved appetite.*[4]

In other species including ours, which are called iteroparous species, individuals do not die immediately but can reproduce several times. They also undergo senescence but of a less violent and more progressive kind.

Some species such as alfafas can display both types of cycle. Alfafas such as the ones around the Mediterranean are annuals whereas those in Northern Europe are perennials and iteroparous. If we prevent the annual species from producing seeds, for instance by cutting their flowers, the plants can survive for several years; usually they die within a few months.

This would seem to prove that death is linked to reproduction (of genes, of course) and that it is genetically programmed. Up until now, we have considered that genes used the individual to reproduce and that, in so doing, endowed it with desirable properties such as better adaptation to the environment, improved survival, *etc*. How can evolution have made genes program the death of the "vehicles" (Dawkins' word) they have procured for themselves?

## Senescence and death

Three hypotheses have been put forward to explain the genetic determinism of senescence:

1. In the first hypothesis, by J.S.B. Haldane, senescence arises from the accumulation in a species of mutations with delayed deleterious effects. The later a deleterious gene acts in an individual's life, the weaker the selection force that eliminates it. It is a quasi-neutral gene. A gene that meant certain death at 500 years of age in humans would not be counterselected even if we did not age and our reproductive activity went on indefinitely because the chances are that we would meet an accidental death beforehand.

It is obvious that once an individual has offspring, he is no serious loss to genes. We can thus model, starting from a non-senescent cycle, how genes causing the loss of some function or other late in an organism's life, appear and become fixed in a species. All these genes together would lead to senescence and finally death.

2. The second hypothesis is by William D. Hamilton who noted that we could turn this argument round the other way. Genes with beneficial effects acting early in an organism's life have the edge over genes with the same effect later on. Natural selection would have gradually improved the performance of young individuals and ignored older ones. This hypothesis cannot account for all aspects of senescence. We would have to suppose that the individuals of the ancestral species looked like old people do today! Its principal interest was that it led to the third hypothesis.

3. Georges C. Williams proposed a synthesis of the two existing hypotheses. In his mind, senescence is explained by the popular adage "you get nothing for nothing"; it is a trade-off as already noted for male sterility.[5] The hypothesis states that improved performance in the young is reflected in lesser longevity. We can find many analogies. For instance, does not a proverb say: "He who wishes to travel far has to spare his horse"? In their race to reproduce, genes would have evolved to build increasingly high performance vehicles at the expense of longevity.

To test these three hypotheses, Michael Rose and Brian Charlesworth[6] selected samples of adult fruit flies (*Drosophila melanogaster*) using criteria of longevity or reproductive ability. They showed that selection in favour of increased longevity led to decreased fecundity at earlier ages, confirming Williams' hypothesis. Genes thus have several types of action during an organism's life.

We might expect all species to be semelparous; why is this not so? Two possible explanations why semelparity may not be dominant are as follows:

Ecological conditions may be influential. Species that specialise in colonising empty spaces have often chosen to be semelparous even though later they are eliminated by less "fast" but better "implanted" species. Let us imagine an open space left barren by a fire. The first settler plants are adapted to this kind of environment (they are sometimes called r strategy species). It is in their interest to produce as many seeds as possible even at the expense of the survival of the individual because, in an empty space, each seed is likely to give rise to a plant. This is semelparity typified. On the other hand, the seeds of a plant in a closed, complex and diverse environment, such as a forest, are less likely to develop because of greater competition. Once a plant has managed to establish itself somewhere, it is in the interest of its genes to stay put as long as possible. Iteroparity is thus preferred.

If external causes of death provoke an individual's death early on before reproduction, selection will favour iteroparous cycles. It is in the individual's interest, once it is over the high risk period, to be able to reproduce several times. On the contrary, if death occurs after the individual has reproduced, the genes have nothing to gain from prolonging its existence by commanding reserves, for instance. Selection will urge individuals to invest their resources in reproduction and keep nothing for later. This probably explains the behaviour of the alfafas mentioned earlier. Alfafas reproduce in spring. The Mediterranean varieties often die of drought in the summer but, having already reproduced, they can be semelparous. On the other hand, in Northern Europe, the young plants have a hard time getting through the winter and, because they reproduce when winter is over, they are iteroparous.

Viewed in the light of neo-Darwinian theory, senescence and death appear as a compromise between the potential usefulness of keeping going and the immediate advantage of reproducing (usefulness and advantage refer to the genes).

**Population dynamics**

The simplest model for the dynamics of a population depends on two parameters:

- the intrinsic growth rate (r) which gives the instantaneous change in the number of individuals in a population that can expand without limit. The growth of such a population is exponential.

- the carrying capacity of the natural environment (K) which sets an absolute limit on population size.

The dynamics of a population is thus a compromise between its inherent and material ability to expand. In 1963 and 1967, the ornithologist Robert H. MacArthur and the entomologist Edward O. Wilson proposed two extreme cases in their theory of dynamic equilibrium: the r strategy where the population multiplies rapidly with no size control; the K strategy where the population takes care not to overload the ecological niche.

## 2. WHAT USE IS SEX?

The tale of the courtship and demise of the praying mantis could well have been an introduction to the second part of this chapter on sex and its evolutionary history. Sex must be really important in life for copulatory habits of the likes of those of the praying mantis to emerge and last! We have postulated that death is linked to reproduction. To detail and defend this postulate, we need to acquire a better understanding of the origin of sex and of the issues at stake. Darwin had already put his mind to the matter. Alert to Malthus' ideas on demography, he tried to apply them to the animal and plant kingdoms. He was already aware of the key role of reproduction when writing *The Origin of Species*:

*A struggle for existence inevitably follows from the high rate at which all organic beings tend to increase.*[7]

### Why take an interest in sex?

What Darwin called "multiplication" is what we usually call "reproduction". Reproduction was defined by Buffon as:

*that property common to man and plants, that power to produce a creature similar to oneself, that chain of individual existences that make up the existence of the species.*

To survive as an individual and to perpetuate itself as a species are the two facets of living material's fundamental reproductive ability. Reproduction is much more than just a property of life distinguishing the animate from the inanimate world; it is a necessity of life.[8,9] Each individual may be subject to wear and tear and destruction, but life can spread, take over new environments, or simply perpetuate itself thanks to reproduction. There is no lack of ways of reproducing whether for macromolecules or higher vertebrates: nucleic acid duplication, cell division, multiplication of multicellular beings by asexual or sexual reproduction, *etc.*

We referred to the link Buffon made between individual and species intentionally because he used it to define species. Before Buffon, the species was an arbitrary level of classification just like order, genus and race. Ever since Buffon, its definition has rested on the criterion of interfecundity and fertility. Populations belonged to the same species if they could produce fertile hybrids by mating. Sex thus provided "*scientific and experimental criteria for defining species that were lacking in all other partitions of the classification of living things*".[10] In the absence of sex as in bacteria, species are less clearly differentiated. None the less, as shown by Miroslav Radman and Christiane Rayssiguier, *Escherichia coli* bacteria and salmonella are not only isolated by a restriction enzyme system but also by reproductive processes.[11]

Reproduction helps us establish another link between individual and species that is subservient to the question "who benefits from reproductive behaviour?" This question has many answers. Reproduction has sometimes been viewed as a luxury physiological function in an individual who survives without reproducing when it is in fact a fundamental biological function for the life of the species. A more moderate view is that reproduction is an essential physiological function integrated with other important functions of the organism. For instance, it determines an individual's biological cycle:

> *In a living being, everything is organised with a view to reproduction. Of what better fate can a bacteria, amoeba, or fern dream than that of forming two bacteria, two amoebae or two ferns?*[12]

Yet another answer is provided by our chosen paradigm summarised in R. Dawkins' words at the beginning of this chapter. Organisms (with their reproductive organs and behavioural patterns) are contingent artifices invented by genes for replication. There is no mention of species, their perpetuation, and expansion. We therefore need to specify under what conditions individuals and species are linked through reproduction. In particular, does their taxonomic relationship have any evolutionary reality? Is the idea that reproduction exists as much for the benefit of the individual (or rather his genes) as of the species a valid explanation, or is it merely an *a posteriori*, mistaken interpretation of a useful classification? Defining species as an ensemble of interfecund individuals might indeed equate reproduction of the individual and of the species but, as we have seen in the previous chapters, a selection process that promotes individual reproduction does not always lead to the best overall gain for all individuals. It seems that reproduction is a highly suitable field for testing the explanatory claims of evolutionary genetics.

Nature has been most inventive in the field of reproduction and, among the many reproductive systems it offers, sex has really proven its worth. More than 95 percent of living species reproduce sexually. For a time, sexual reproduction was not thought to take place in lower order forms but we now know of many cases of parasexuality, *e.g.* conjugation in ciliates (unicellular animals - the best known is the paramecium - which cling together by their mouths, exchange their nuclei, and then separate), hologamy in *Chlamydomonas* (a unicellular algae where the two gametes that merge are indistinct from normal individuals), or somatogamy in *Basidiomycetes* (mushrooms in which normal somatic cells mate, the nuclei merge later).

But what is sex? In sexual reproduction, descendants are produced by the merging of two sex cells (or gametes) from separate individuals. The resultant egg has two homologous batches of chromosomes that are not absolutely identical; one comes from the father, the other from the mother. In adulthood, the individual born from this egg passes on to each of its gametes, during meiosis, only half of its chromosomes but not as an original parental batch but in a random mix. The distribution does not involve homologous chromosomes from the two parents; there are exchanges called crossing-overs. Both these occurrences - random distribution and crossing-over - seem to have a single objective, to produce new genetic combinations of the paternal and maternal genomes. Thus, each zygote lies at the origin of a unique genetic entity (that can consist of several individuals: twins, grafts, clones, *etc.*).

## At the beginning there was information exchange

How did sex come about? There is no reason to believe that mitosis or cloning preceded meiosis and its recombinations. On the contrary, as we shall see, genetic recombination probably quickly became a repair mechanism for the genetic information borne by the first molecules of life. The origin of life itself is still unknown but we must make room for at least one exception to Pasteur's rule of no spontaneous generation. Life probably emerged from, and at the expense of, the mineral world. This was also Darwin's view. All that was needed was a single exception to the rule. This would agree with the hypothesis that life has a single origin because of its molecular unity. None the less, we cannot exclude the possibility that other living forms emerged later but were immediately eliminated by pre-existing forms. For quite some time, the gap between the mineral world and the fully formed cell seemed unbridgeable but, nowadays, we can conjure up the intermediate cell-free steps because "*we can use perfectly abiotic molecules to perform specific catalytic reactions that were earlier thought to be the sole preserve of enzymes extracted from living organisms. In plain, selectivity and catalysis could have appeared at the dawn of life outside living cells*".[13] However, if these macromolecules were to give rise to life as we know it today, they had to fulfil two conditions: (1) reproduce and replicate hereditary material to provide a virtually identical copy and (2) establish a code of correspondence between nucleic acids and proteins. There was also the issue of enzymes for making proteins from nucleic acids. The discovery of a self-excising ribosomal RNA that needs no enzyme raised the chicken and egg question again. Which came first, protein or RNA? And what about co-evolution?

In the primeval prebiotic soup of three to four thousand million years ago, the genetic information contained in the first DNA and RNA sequences (which we assume could replicate) came up against two hurdles. First, information deteriorated fast because of mutations due to radiation falling on the Earth's surface and because of the inefficiency of the first replication and DNA repair systems. Second, in the absence of true cells, the integrity of the "organisms" was not safeguarded. Organisms were poorly isolated from the outside so that spontaneous recombination of many different pieces of information could have occurred quite easily.

Paradoxically, at this stage of evolution, these two obstacles probably balanced each other out and helped promote the development of life. When information became useless because of a mutation, the "individual" could exchange part of its genome with that of a "partner" which had not lost the same piece of information. To use a metaphor, with two copies of a text that have been poorly transcribed and with a little bit of luck, we can work out what was the original text. From the two statements "This sentence is +$-£*%@#" and "This pzzzze is correct", we can obtain "This sentence is correct" by combining the words before and after "is". From the outset selection must have perpetuated at least some of these spontaneous exchanges. Most research-workers support this hypothesis today. To test it, R. Michod of the University of Arizona submitted bacteria of the species *Bacillus subtilis* to mutagenic radiation under three conditions: in the presence of foreign DNA (which they can take up) in the culture medium before and after irradiation, and in the absence of foreign DNA.[14] The bacteria that received foreign DNA after irradiation survived best because integration of this DNA after mutation helped compensate for the damage undergone by the bacterial DNA. It is for such reasons that sex, which allows information exchange through meiosis, was selected at life's beginnings.

Another non-exclusive hypothesis for the origin of sex is based on the behaviour of existing plasmids (DNA molecules that have a certain autonomy of replication) and viruses. In order to reproduce, they go from bacteria to bacteria at times taking along with them part or most of the chromosome of the bacteria they have just quit. The transfer of genetic information from one entity to another - leading to sexual reproduction -

would have found its origin in the infectious nature of these plasmids and viruses.

As evolution progressed, DNA replication probably became increasingly faithful and the sequences were kept ever more rigorously apart. When the cell began to evolve, the chromosomes of bacteria and other prokaryotes (organisms without nuclei) were located in the cytoplasm and not isolated in any way. A DNA fragment that penetrated a cell could easily insert itself into the chromosome and so to speak "infect" it, as some viruses that parasite bacteria still do. The function of bacterial restriction enzymes is to cut up all foreign DNA into little pieces and so hinder the intruders' actions. As time went by, natural selection would have perpetuated the bacteria that were best able to destroy the intruding parasite DNA.

The cells of eukaryote organisms (animals, plants, fungi, yeasts) have a nucleus; a nuclear membrane isolates the chromosomes from the surrounding cytoplasm. To reach the chromosomes, a DNA sequence entering a eukaryote cell must cross an additional barrier. Restriction enzymes and the nucleus are thus means of isolating and protecting the genome. However, isolation needs to be accompanied by improved DNA replication since otherwise mounting errors at each generation could end up by killing lineages overprotected from spontaneous recombinations.

Sexual reproduction is thus a compromise between isolation and exchange of genetic information. Exchanging too much is detrimental to creating an "identity", too much isolation means succumbing to mutations. Selection has gone even further by favouring those organisms that can control this process and choose their partners. This is the origin of meiosis and of fecundation in eukaryotes. An even higher level of complexity is encountered in the sexual behaviour of higher-order animals. The mechanisms of choosing a partner are at the origin of establishing species, if we stick to our definition of a species as a set of individuals that can exchange genetic information.

In the beginning, cloning did not reign alone; recombination played a key role. Sex, the modern version of archaic exchanges, still prevails. We now have to find out why.

## Why sex?

As we have already noted, currently there are few asexual species. They belong to groups as diverse as plants, insects, lizards or fish[15] and reproduce using a wide variety of systems, with or without the production of sex cells.

An unfertilised female can produce a descendant genetically similar to herself using her own sex organs. This type of reproduction is called parthogenesis; the green-fly is a well-known example. Parthogenetic fish such as *Poeciliopsis monacha* insist upon male intervention but the male belongs to another species since parthogenesis - the only way these fishes can reproduce - produces females only. The chromosomes of the male are not passed on to the descendants. Parthogenesis can occur by automixis when two sex cells of the same female merge. This mode of reproduction is encountered in some insects (fruit-flies, butterflies, *etc.*), fish (*e.g.* crucian) and lizards. Whip-tailed lizards, among them *Cnemidophorus uniparens*, are all females that reproduce by automixis. They make a pretence of mating, adopting a posture very similar to that of closely related ancestral sexual species. Another reproduction mode is producing an egg directly by normal division without meiosis and thus with no change in chromosome number, as in some reptiles, fish and plants. The plants produce pollen even though they have no use for it. Finally, autogamy, as practised by hermaphrodites, is a way of losing sexual reproduction. Systematic self-fertilisation is practised by some snails and many plants.

In all of the above instances, there is loss of sexual reproduction, never a lack in its acquisition because, in each case, we can find traces of sex in the past. Support for this insight comes from the presence of asexual species in all major living groups. Sex would be an ancestral character that has become less necessary for the short-term function of the species because current replication systems are reliable. Is it then an archaic relic, or a practice that is profitable in the long-term? The usual explanations for the perpetuation of sex all revolve around the same central idea that recombination is a source of genetic variation. According to Darwin, evolution is the resultant of two factors, hereditary variety and natural selection. Which are then the processes that generate variation? Mutation (*i.e.*, a random change in genetic information that can be passed on to offspring) and genetic recombination (a random process that concerns all or part of chromosomes). They are not independent processes since recombination ensures the spread of new information produced by mutation. High variability enables rapid evolution when the environment changes because natural selection then has a wide choice. This was the reasoning on which Hermann Muller's hypothesis was based. He postulated in 1932 that, for an organism to adapt well to an ever-changing environment, genes must be able to mutate and that there is therefore an advantage to sexual reproduction.

In a population that does not reproduce through sex, favourable mutations can be incorporated into one and the same individual as long as they occur in succession in the same lineage. This needs time (at least N generations for N mutated characters) and enough carriers of the first mutation for the second mutation to be able to emerge in them (which further increases the time needed). On the other hand, in a sexed population, sex ensures that genomes are continually mixed. Two favourable mutations, each in a different lineage, can unite quickly without needing to be present in large numbers of individuals.

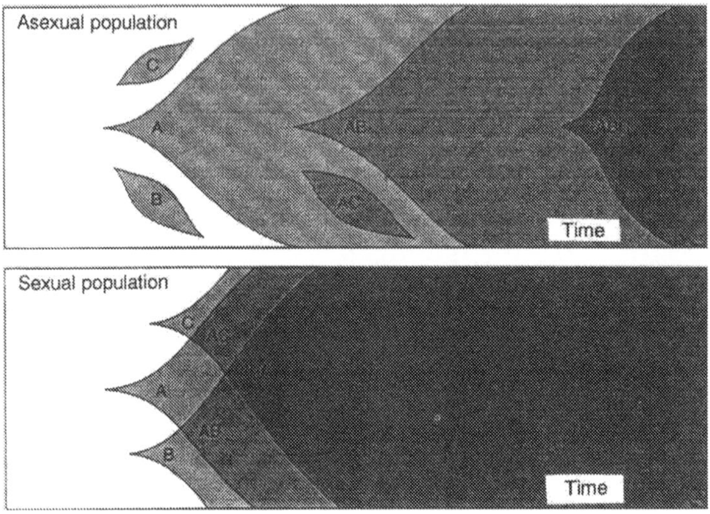


*Figure 6.1.* In a constantly changing environment, sexual reproduction is advantageous. This is because genomes are constantly shuffled; the traits needed to face the new environmental conditions are thus more likely to co-occur in the same individual. Let us suppose that for an organism to adapt well to its environment at a given time, it needs three genes, A, B and C, resulting from mutations. For an asexual population it takes a long time to fix these genes. We have to wait for the second favourable mutation to arise in an individual who underwent the first mutation, then for the third mutation to occur in the double mutant (top panel). On the other hand, with sex, the three mutations that have appeared in different lineages can combine rapidly (bottom panel). (Adapted from Muller, H.J., 1932, Some genetic aspects of sex[16]).

The greater the number of advantageous mutations in a population, the greater the benefit of recombination; all that has to be done is to combine these mutations in a single individual. Large populations are thus at an advantage. In 1965, J.F. Crow and Motoo Kimura showed that recombination hardly benefits small populations because of the time required for the other mutations to crop up. The rate at which mutations appear is too low compared to the time taken for each mutant to fix. Moreover, only large populations can retain genuine polymorphism. And lastly, sexual reproduction only surpasses asexual reproduction if emergent mutations take time to establish and are not overtly advantaged; we refer to a low effect mutation.

Yet another advantage to recombination - to be compared with the combining of favourable mutations - is avoidance of the accumulation of unfavourable mutations. This always occurs in asexual reproduction. In 1964, Muller compared such accumulation to the way a ratchet works.[17]

A ratchet is a set of teeth on the edge of a rack in which a pawl prevents backward motion (as in a hand brake or watch spring). In a sexual species, two genomes carrying different deleterious mutations can recombine to produce a genome with no mutation. Because this is not possible in an asexual species, the minimal number of mutations per genome undergoes a constant increase with time. This is analogous to a ratchet that can only move forward.

**Mutations and populations**

Let's take a population of haploid individuals of finite size N, at different generations. The individuals' genome is made up of L loci (where L is large) that code for essential proteins with different functions. Because of replication errors, functional alleles may become deleterious alleles with a probability (or deleterious mutation rate) of u, which is considered to be rather low. Deleterious alleles are counterselected. Carrying a deleterious allele reduces an individual's fitness by a fraction s (its fitness becomes 1 - s compared to 1 for individuals with no mutation). This deleterious effect is assumed to be multiplicative, in other words, carrying k deleterious mutations reduces fitness to $(1\text{-}s)^k$. The more mutations an individual carries, the less it multiplies. The status of the population is defined by the number of individuals with 0, 1,......i,....,L mutations. Two opposing forces influence this distribution: mutation pushes individuals of one category into a higher category (from i to i+1); selection advantages the reproduction of individuals with few mutations (it increases the frequency of category 0 with respect to category 1, and so on). For an infinite population, an equilibrium is reached. On the other hand, for a finite population, the numbers in each category undergo random variations (because some individuals multiply and others do not). These variations only count for category 0 because, should its numbers fall to 0, the category cannot be recreated from other individuals (reverse mutations from a deleterious allele to a functional allele are extremely rare). The distribution is thus shifted by one notch and the "ratchet" prevents it from going backwards (line 2 of Fig. 6.2). However, the distribution reaches a new state of equilibrium where category 0 is henceforth empty. Random variations again come into play and end up by emptying the category of individuals with just one mutation which cannot be recreated either (line 3). The population will thus accumulate a large number of mutations by a series of shifts in the different categories. Mathematical analysis indicates that the ratchet works when the probability of "emptying" the category with the least mutations is high, that is when u >1- s or, in biological terms, when the mutation rate per genome is greater than the disadvantage due to a mutation. For u = 1-s, the ratchet works for a number of generations below 1000.

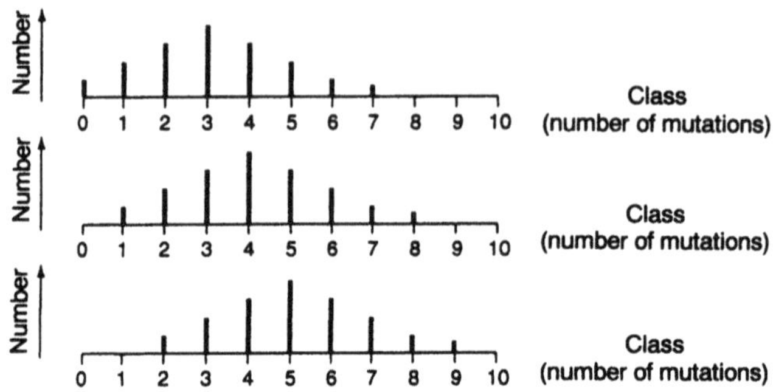


*Figure 6.2.* Muller's ratchet hypothesis.

How can we escape from Muller's ratchet? How can we make sure that the first category - unmutated genomes - never dies out? The answer is by having a large number of copies, a low mutation rate and a small number of loci. None of these conditions apply to species with a large genome but small in numbers (such as Man). Sexual reproduction, which combines intact chromosome regions in a single genome, thus comes across as a special kind of DNA repair process. Interestingly, what we commonly call repair in molecular biology (correcting inappropriate pairing, deleting nucleotide dimers, *etc.*) is carried out by the same machinery as does the recombining. Repair might be an early function of recombination, and thus of sex.

## Why not sex?

The above view is common among biologists and often reported by the popular scientific press. Sexual reproduction creates diversity, and this diversity may help species overcome future obstacles. Moreover, sex sort of repairs the genome of living beings undergoing mutations. However, are there benefits for a species that are not long term?

Muller's 1932 hypothesis[16] of the coming together of beneficial pieces of information is only relevant in an environment undergoing constant change. For a mechanism like sex to be perpetuated in a fairly stable environment, there have to be occasional brutal changes - "catastrophes" - to which only some individuals are able to adapt. A species whose members are diverse would have the best chances of survival. But is it not at risk of losing sex whilst the environment is constant?

Muller's 1964 hypothesis on ratchets and sex as a repair job applies to the invention rather than perpetuation of sex (in the former case, highly exposed gene sequences need repair). Is this not a Panglossian situation (see next chapter) where "all is for the best in the best of worlds" is translated into "sex exists for the future adaptation of the species"? Explaining sex maintenance in this way confers natural selection a kind of "prescience". It is asked to retain a strategy that might provide future dividends but that is extremely costly in the short term. Support for such reasoning would come from a phenomenon that counterselects sex and opposes the above examples "in favour of" sex. It exists, and has been called the cost of meiosis and the cost of males by John-Maynard Smith and Georges C. Williams (in 1971 and 1973 respectively). Why? Because the female gamete which is much larger than the male gamete (this is how they are defined), as well as the female herself in higher-order animals, provide most and sometimes all of the energy that is needed to produce the egg, to help the embryo grow (there may be enough reserves in the female gamete as in the vitellin sac of young fish), and to feed the embryo. The male may invest substantially in seeking out a partner and mating but, in most cases, he is just a provider of sperm. If the female were able to reproduce on her own, there would be no cost to producing a male. The genetic information that determines females makes them remarkable reproduction tools (an echo of one of our earlier conclusions). However, before producing a new organism, these genes have to be halved by cell division (meiosis), the lost half being replaced by genes from the male. If the female cell had just undergone normal division (mitosis) and had not been fertilised, each of its genes would have been reproduced twice as many times (compared to its average reproduction rate over a large number of meioses). In other words, asexual behaviour would be twice as beneficial as sexual behaviour. We say that the cost of meiosis per generation is equal to two in the female. How do we then explain that selection acting on genes has perpetuated a mechanism (meiosis), which, even if it is beneficial in the long term, halves the speed of reproduction of the genetic information that determines it?

The same question arises when we compare the fates of genetic information carried by a parthenogenetic lineage and a sexual lineage. If, for example, each female has exactly two offspring, we obtain two females in the parthenogenetic lineage and one male plus one female in the sexual lineage. At the next generation, the size of the sexual family will be unchanged but that of the parthogenetic family will have doubled. Once again, the cost of meiosis is a factor of two. After 10 generations, the parthogenetic lineage will be 1000 times larger than the sexual lineage and will have virtually eliminated it, in the absence of other factors. The long-term advantages of sexual reproduction would not have had time to show up.

These two examples thus suggest that there is an important short-term advantage (of the order of a factor of two at each generation) of sexual reproduction over asexual reproduction.

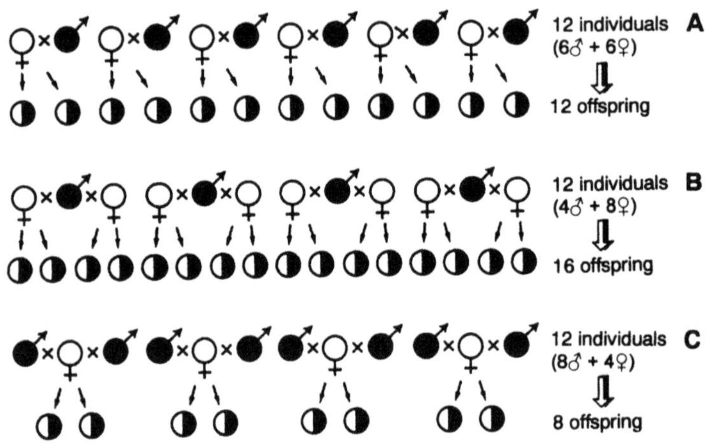


*Figure 6.3.* The cost of producing males. Let's consider a population of 12 individuals in which the fertilised females produce a set number of two offspring. A balanced sex ratio leads to stable population numbers (A), females in excess increase population size (B), and males in excess decrease it (C). Notwithstanding, our conclusion in Chapter 5 is still valid: The individuals of the rarer sex are those that pass on their genes best.

## Sex as a short-term advantage

The need to find short-term benefits to the perpetuation of sex has given rise to several hypotheses but few are satisfying. A first model was proposed by Georges C. Williams in the 1970s. His balance argument is based on the idea that, in species where sexual and asexual reproduction coexist, some mechanism perpetuates sex despite its disadvantage. Williams' model, by making many assumptions, gets several parthogenetic generations to alternate with one sexual generation. If the offspring are all in an environment unknown to the mother, it is advantageous to use recombination to generate the variety lost during the asexual phase despite the cost of meiosis. The number of asexual generations required to make one sexual generation "pay off" then depends on the number of offspring per individual; sex pays off only when this number is high enough. G. Williams calculated that to do away completely with asexual reproduction, each individual must have about several hundred million offspring. This hypothesis is inapplicable to most species even if the reproduction rate of their distant ancestors is unknown.

Williams' unrealistic hypothesis is of historic interest only. A single later hypothesis, by J.A. Antonovics and William D. Hamilton, has resisted the test of time but even it is not universally accepted. The hypothesis is based on the idea that an individual has to adapt daily to the organisms around it (pathogens, parasites, predators, competing species). It postulates that host-pathogen relations, competitive interactions, and perhaps predators force species to evolve rapidly. Gene recombination, which redistributes genetic information, is needed to create the high diversity that permits more efficient selection. This hypothesis is equivalent to the Red Queen hypothesis, on a much shorter time-scale.[18]

Although theories accounting for the perpetuation of sex are scarce, experimental studies have brought to light a network of interknit factors that might do so. The assumption that the principal reason for sexual reproduction is the benefit obtained from mixing genes gives way to a hotch-potch of reasons at times only distantly related to sexual reproduction. We shall distinguish between two types of constraints: *ad hoc* developmental constraints and bizarre physiological or genetic constraints.

### *Ad hoc* developmental constraints (or when there will no longer be any men...)

*Ad hoc* developmental constraints are by far the more common and varied. They range from the biological incapacity for an egg to develop without gametes being produced, combined with the intriguing need for sexual reproduction to produce certain structures. In mammals, no offspring can be produced by parthogenesis, that is without gametes and meiosis, although many research-workers have attempted to do so. (In 1997, an offspring was produced asexually in a mammal but it is most unlikely that the series of events leading to the birth of the famous sheep Dolly would occur in nature[19]). In a French television programme in October 1986, the consequences of having children of purely maternal origin, conceived without sperm, were discussed. The programme was based on P. Soupart's work in the States in 1977 on the development of mouse embryos from two combined female sex cells (ovocytes). Were baby mice really born without a father? P. Soupart did not publish his findings and, apparently, only spoke of them in private. Moreover, his baby mice had the misfortune of being eaten up by their mother! Like the parthenogenetic rabbits a scientist claimed to have produced at the beginning of the 20th century, these mice probably belong to the vast repertoire of scientific jokes. Subsequent work did not reproduce Soupart's results but, on the contrary, suggested that a male gamete is essential for the development of a mammal embryo to term. Thus, even if parthogenesis does occur in most animal groups, it is notably absent in mammals. No one has ever observed the birth of a mammalian parthenogen.

Parthogenetic mammalian embryos can, however, be obtained experimentally by activating ovocytes with chemical or physical agents. There is even a strain of mice, called LT/s mice, that produces a high natural rate of spontaneously activating ovocytes. Embryonic development starts off "normally" and implantation into the uterus of a recipient mouse is possible, but these embryos never survive beyond the halfway stage of gestation. The parthenogenetic cells are nevertheless viable. When combined with normal cells to form chimeric embryos, they do not prevent the baby mouse from developing to term. They differentiate in all tissues as long as they do not account for more than 20 percent of the cells. Apparently, these parthenote cells lack "signals" required in morphogenesis.

Rapid progress in microsurgery techniques since the 1970s enabled removal of one of the two nuclei (or pronuclei), whether male or female, just after fertilisation of the ovule. Only genetic information, and no other potentially influential material such as the membrane and cytoplasm of the spermatozoa, was removed in this way. The procedure yielded gynogenotes (the genetic material is of strictly female origin but there is an extranuclear contribution from sperm) and androgenotes (the genetic origin of the embryo is entirely male). Both gynogenote and androgenote embryos developed just like parthenogens. Development was arrested at or just after implantation. These results confirmed that genetic material from both mother and father is needed for the embryo to reach term. Moreover, the two pronuclei are not interchangeable and thus not equivalent. The egg recognises the origin of all genetic material. Yet, in theory, the genome is virtually identical in both sexes, at least in consanguinous lineages. Does this mean that the two genomes were "marked" differently?

According to Azim Surani and his colleagues at the Institute of Animal Physiology in Cambridge, UK, gene imprinting endows elements of the reproductive cell's hereditary material with a kind of memory of their parental origin. A batch of chromosomes marked father and a batch marked mother may have to combine for the embryo to develop normally to full term. The imprint would be extra information carried by genes inherited from parents. What might be the mechanism? Azim Surani and Wolf Reik have proposed a hypothesis that supports the work of Carmen Sapienza's team at the Ludwig Institute on Cancer Research. The DNA would be methylated (addition of a methyl ($CH_3$) group to the DNA molecule).[20]

The hypothesis was tested by transgenesis. An individual's genome was artificially modified by injection of a foreign gene (transgene) into the embryo. Methylation of the transgene was measured in the DNA of offspring from crosses between a transgenic male (carrying the transgene) and a wild-type female (non transgenic), and between a transgenic female and a wild-type male. A series of crosses made the transgene alternate between the male and female. It remained marked but its methylation level, which depends on the DNA's parental origin, reversed each time it passed into the other sex. We shall not delve into all the experimental results[21] but note that low-level methylation can be an attribute of either the female or male lineage. In the male, it would occur in the testes or sperm. Moreover, although the transgene can insert itself anywhere in the genome, it does not undergo differential methylation in all lineages; thus only certain chromosome regions are imprinted.

Such marking can be explained by the distribution of resources between the female and her embryos, as outlined in Chapter 5. The paternal genes have all to gain from requisitioning maternal reserves for the embryo that carries them. On the other hand, the female must fight this and spare herself for future embryos. This is confirmed by experiments in which two male or female gametes are united. In the first case, the vitellin stocks are high but the embryo does not develop; in the second, the embryo develops but the vitellin shortage causes death. In species where resources are allocated after fertilisation, genes from each sex lineage must unite to reach a compromise. Because the male is an absolute requirement in mammals, parthogenesis is not possible.

Greenfly (aphids, *Homoptera*) are a rather more complicated case. They can reproduce by parthogenesis but cannot produce eggs in this way. Parthogenesis is strictly viviparous. Because the egg is the only form that withstands the cold, greenfly in temperate climates have to reproduce sexually at least once a year, in the autumn, in order to survive. When the winters are mild, strictly parthogenetic lineages invade the population. After a bitter winter, only the sexual lineages leave offspring.[22] Greenfly have thus retained sex to withstand the cold. It is a short-term advantage of sex that has nothing to do with gene shuffling.

Similarly, many plants produce structures to be dispersed (seeds, spores) by just the sexual route; asexual reproduction organs (bulbils, runners) tend to stay put. The reason why selection maintains the sexual structures is not an advantage due to gene shuffling but rather the ability to colonise new territories.

## Bizarre physiological or genetic constraints

Physiological or genetic constraints could also help explain sex maintenance. In Hawaii, there are parthogenetic lineages among fruit flies such as the *Drosophila mercatorum*. When, in 1982, Alan R. Templeton of Washington University Saint-Louis made them compete with sexual lineages, the sexual lineages won. Apparently, the way parthogenesis works in these insects induces dysfunctions in the offspring.

What about plants? Most are hermaphrodites. Self-fertilisation reduces the cost of producing males since the two chromosome batches come from the mother plant. Here again, cross-fertilisation (and therefore sex) often seems to be maintained because of a wide variety of constraints. In groundsel (*Senecio sp.*), each individual carries next to flowers that self-fertilise flowers that do not; sex is thus perpetuated. In the 1980s, R. Abott showed that self-fertilising flowers are less fertile than other flowers even when out-crossed. This is due to the flower's structure and is independent of reproduction mode. Here, the constraint that perpetuates sex is the inability to produce autogamous flowers with normal fertility whereas natural selection favours maximal fertility. As for thyme (*Thymus vulgaris L.*), cross-fertilisation is maintained by females being next to self-fertilising hermaphrodites. Females are not maintained because of any adaptive advantage for the species but because complex genetic and evolutionary interactions exist between nuclear and cytoplasmic genes.

At first sight, the conclusion of our overview is not very satisfying. The theoretical approaches have shortcomings. We are faced with a host of species-specific cases where sex is perpetuated because of constraints unrelated to what sexual reproduction is really about (gene shuffling). Might there not be a single cause but many independent causes for something as widespread as sex? Many people object to this paradoxical idea, preferring to believe in an unknown factor that is perhaps linked to the evolutionary race between organisms and parasites. But is there not here an epistemological misunderstanding about causality? Several independent causes can have the same effect - i.e., maintain sex - if there is a higher-order constraint acting on the species.

To address this point, we need to view the situation from the angle of ranked selection units as discussed earlier. In the case of the evolution of sex, the avatars must be ranked and there must also be a "chronological" ranking. (Avatar is defined in Chapter 5. An avatar passes on genetic information and is its material support). There are indeed many short-term reasons for keeping sex. These fall into two categories: constraints (or the space of what is possible) and individual selection which concerns the survival of genetic variation when it appears. It is among the species that have short-term reasons to reproduce sexually (resistance to the cold, dispersal of the species, *etc.*) that we shall find those that will survive in the long term and adapt best to changes in the environment. The long-term need for sex acts as a higher integration level on species. The long-term advantage due to sex identifies, amidst all the forms that see the light of day, those that cannot lose sex because of a specific constraint.

In the end, the two questions "Why do species in general reproduce sexually?" and "Why such and such a species reproduces sexually?" yield two different answers. The first question addresses a general and slow mechanism; the second deals with a variety of fast-acting mechanisms. Thus, even if we know why each species reproduces sexually, we cannot know why most of them do. Understanding the evolution of sexual reproduction in the living world is knowing all the short term mechanisms that force existing species to be and remain sexed. It is also knowing the slowly acting mechanisms at a higher integration level that have led to most species, in whom sex is not a constraint, to become extinct. In other words, if sexual reproduction is so widespread, it is because the species that one day had the "misfortune" of being able to do without sex did not make any old bones.

**NOTES**

1. Burnet, M., 1982, *Le Programme et l'Erreur. Une Histoire Naturelle de la Mort et de l'Hérédité*, Series: Sciences d'Aujourd'hui, Albin Michel, Paris, p. 30.
2. Hamburger, J., Postface to Burnet, M., *Le Programme et l'Erreur*, p. 352.
3. Vincent, J.D., 1996, *La Chair et le Diable*, Odile Jacob, Paris, pp. 132-134.
4. Fabre, J.H., 1897, *Souvenirs Entomologiques. Etudes sur l'Instinct et les Moeurs des Insectes*, Cinquième série: *La Mante. Les Amours*, Delagrave, Paris, pp. 306-307.
5. See Chapter 4.
6. Rose, M., and Charlesworth, B., 1980, A test of evolutionary theories of senescence, *Nature* 287: 141-142.
7. Darwin, C., 1859, *On the Origin of Species by Means of Natural Selection*, Murray, London.
8. Williams, G.C., 1975, *Sex and Evolution*, Princeton University Press, Princeton.
9. Maynard-Smith, J., 1978, *The Evolution of Sex*, Cambridge University Press, Cambridge.
10. Langaney, A., Pellegrini, B., and Poloni, E., 1989, L'homme descend du sexe, *La Recherche* 20 (N° 213): 998.
11. Rayssiguier, C., Thaler, D.S., and Radman, M. 1989, The barrier to recombination between *Escherichia coli* and *Salmonella typhimurium* is disrupted in mismatch-repair mutants. *Nature* 342: 396-401.
12. Jacob, F., 1970, *La Logique du Vivant, Une Histoire de l'Hérédité*, Gallimard, Paris, p. 12.
13. Danchin, A., 1988, L'origine de la vie, *La Recherche* 19 (N° 201): 881.
14. Michod, R.E., Wojciechowski, M.F., and Hoelzer, M.A., 1988, DNA repair and the evolution of transformation in the bacterium *Bacillus subtilis*. *Genetics* 118: 31-39.
15. Although it is difficult to count the number of species (as there is no sexual reproduction), these forms nevertheless make up a small percentage of the corresponding genus or family in terms of numbers of individuals.
16. Muller, H.J., 1932, Some genetic aspects of sex, *American Naturalist*, 66: 118-138.
17. Muller, H.J., 1964, The relation of recombination to mutational advance, *Mutation Research*, 1: 2-9.
18. See "Struggle for life and a royal race" in Chapter 7.
19. As a reminder, Dolly was produced from a diploid nucleus of a mammary gland cell that was reimplanted into an egg. It is not therefore a cell that was forced to enter ontogenesis.
20. Among the four bases that make up DNA - adenine, cytosine, guanine and thymine - the only one that can be modified is cytosine. In the sequence cytosine-guanine, it can carry a methyl group.

21. See Tastemain, C., 1987, Les chromosomes se souviennent aussi, *La Recherche* 18 (N°194): 1537-1539.
22. Rispe, C., Pierre, J.S., Simon, J.C., Gouyon, P.H., 1998, Models of sexual and asexual existence in aphids based on constraints. *J. Evol. Biol.* 11: 685-701.

# Chapter 7

# WHAT CAN BE DONE, WHAT CAN'T BE DONE

To biologists and those who seek in evolution a key to the meaning of life, adaptation is a major stumbling block. We tend to consider organisms and organs as the jewels of natural selection. Natural selection has endowed them with optimal adaptation to their environment or function. In this way, we rediscover and safeguard the idea that life's organisation is harmonious. The previous chapters have already given this idea a knock. In this chapter, we shall give it the *"coup de grâce"* by divorcing the natural selection process from such an adaptive view. For this, we must introduce concepts we often avoid, or that are rather *risqué* in their use: chance, contingency, finality.

## 1. A VISIT TO ST. MARK'S

St. Mark's Cathedral in Venice has a very special place among the symbols used by evolutionists. In 1978, at the Royal Society in London, S. J. Gould, speaking in the name of his colleague R.C. Lewontin, started his talk by describing the cathedral's dome. He claimed that the mosaics, rather than being the fruit of a premeditated plan, were the consequences of constraints imposed by the architecture.[1] In an article, the two American biologists went on to describe King's College Chapel in Cambridge where the Tudor Rose and portcullis alternate in great harmony. Here also, they claimed that the surface embellishment, however appropriate and artistic, was a mere by-product of architectural constraints. What were they getting at? Apart from urging tourists to visit these sites!

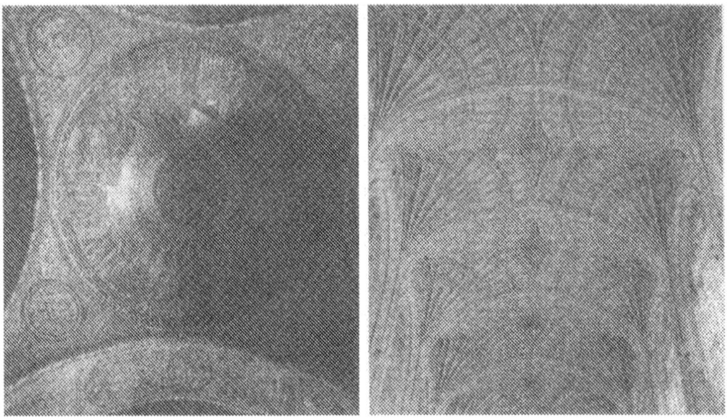

*Figure 7.1.* The cupola with spandrels of St. Mark's (left) and King's College Chapel (right)

**The dome of St. Mark's as seen by Gould and Lewontin**

> *The great dome of St Mark's Cathedral in Venice presents in its mosaic design a detailed iconography expressing the mainstays of Christian faith. Three circles of figures radiate out from a central image of Christ: angels, disciples, virtues. Each circle is divided into quadrants, even though the dome itself is radially symmetrical in structure. Each quadrant meets one of the four spandrels in the arches below the dome. Spandrels - the tapering triangular spaces formed by the intersection of two rounded arches at right angle - are necessary architectural by-products of mounting a dome on rounded arches. Each spandrel contains a design admirably fitted into its tapering space. An evangelist sits in the upper part flanked by the heavenly cities. Below, a man representing one of the four biblical rivers (Tigris, Euphrates, Indus and Nile) pours water from a pitcher into the narrowing space below his feet.*
>
> *The design is so elaborate, harmonious and purposeful that we are tempted to view it as the starting point of any analysis, as the cause in some sense of the surrounding architecture. But this would invert the proper path of analysis? The system begins with an architectural constraint: the necessary four spandrels and their tapering triangular form. They provide a space in which the mosaicists worked; they set the quadripartite symmetry of the dome above.*[2]

According to the two biologists, whoever thought that the primary intention of the architects of St. Mark's and King's College Chapel was to embellish would only be paraphrasing the words of the good Doctor Pangloss, the character created by Voltaire in *Candide*:

> *Pangloss taught metaphysical-theological-cosmological simple-mindedness. He could prove admirably that there was no effect that did not have a cause, and that, in this best of possible worlds, the castle of his lordship the baron was the most beautiful castle and madame the best of possible baronesses.*
>
> *It has been demonstrated that nothing could be different from the way it is, he used to say. Because everything is made for a purpose, it must be made for the best purpose. We should note that noses were made to bear spectacles, and we have spectacles. Legs are clearly for breeches, and we have breeches. Stones were made to be hewn and to build castles. His lordship has a very nice castle; the greatest baron of the land must have the best accommodation. And since pigs are made to be eaten, we eat pork all the year round. It follows that those who have claimed that everything is good have said silly things;* ***they ought to have said everything is for the best.***[3]

According to S.J. Gould and R.C. Lewontin, to claim that the architects' intentions were, first and foremost, to embellish is to use a Panglossian paradigm. By repeating this neologism, they were, in fact, criticising the way evolutionists had overemphasized organism adaptation since the thirties, and were directly attacking neo-Darwinian theory.

The expression "Panglossian" was introduced by J.B.S. Haldane and is now in current use among evolutionists. The term criticises an exaggerated form of finalism at a time when finalism is no longer taboo in biology. It denotes a gene's "will" to replicate. The word "will" is to be construed in the sense given by the German philosopher Arthur Schopenhauer. The will is not specific to humans but a feature of the living world. Little by little, it has produced different kinds of living creatures and, in the end, consciousness which is a human epiphenomenon.[4]

## 2. BRACHIOPOD ARCHITECTURE OR "SEILACHER'S TRIANGLE"

Why shouldn't the living world display architectural constraints similar to those of our buildings? These constraints, rather than adaptation to a particular function or use, might be the "*raisons d'être*" of our organs or structures. R.C. Lewontin and S.J. Gould give several examples; one of them is the brachiopod. Brachiopods are a branch of marine invertebrates similar to bivalvular molluscs such as the clam. The example seems to be particularly dear to S.J. Gould who, besides citing the seminal work of the German palaeontologist Adolf Seilacher, studied brachiopods himself.[5] Seilacher studied the morphology of many shells and showed that a number of traits were not only explained by inheritance, or the need to adapt to special environments, but also by non-adaptive morphogenetic programmes. This applied, for instance, to divaricate structure.

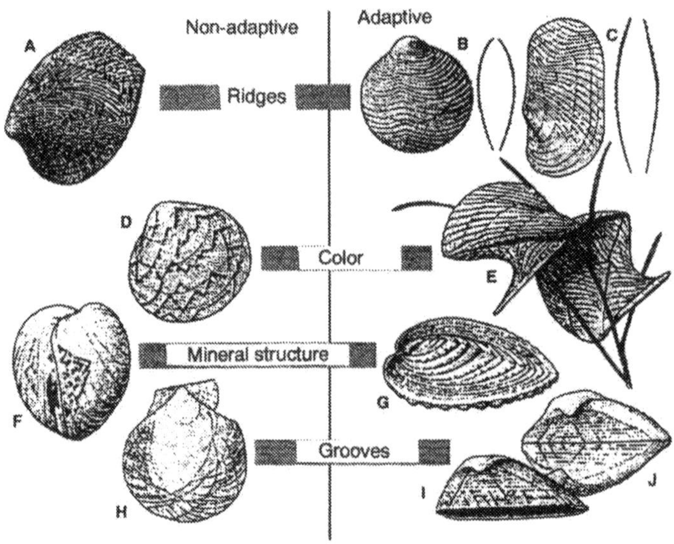


*Figure 7.2.* Divaricate architecture in shells (brachiopods and bivalvular molluscs). The German palaeontologist A. Seilacher showed that many morphological traits in molluscs were not due to adaptation but to a specific mode of construction known as divaricate architecture. In some species, these traits may have acquired adaptive significance later on. Divaricate architecture is seen in ridge or groove morphology, coloured motifs, and mineral structure. In the diagram above, the traits on the left have no adaptive use unlike those on the right. Species A (*Petricola lapicida*) has a herring-bone pattern of ridges of no adaptive value. This is the most common case. Because of their wavy ridges, species B and C can bury themselves in sand or mud. Species D (*Lioconcha castrensis*) derives no benefit from its coloured ridges as it lives underground whereas species E (*Pteria zebra*) uses its coloured ridges to hide in an arborescent colony of Hydrozoa. Species F (*Petricola carcinoides*) and G (*Corculum cardissa*) have triangular shaped divaricate mineral structures. Those of G seem to be of some adaptive value since they are translucent, letting light pass through the shell, so that symbiotic algae living inside can carry out photosynthesis. The genus Arca (I and J), unlike the scallop (H), has divaricate grooves that firmly anchor the ligament that closes the valves. (Adapted from Seilacher, 1972[1]).

In a divaricate shell structure, divergent lines or a herring-bone pattern mark out decorative ribs, coloured motifs, mineralised internal structures, or groove shapes. Seilacher distinguished morphological traits that have adaptive value (the majority) from those that do not. For instance, in the presence of predators, some species use chevron colour as camouflage. But of what use is it to clams which live buried in the sand or are covered by a periostracum that conceals shell colour?

There is a "mechanical" explanation for these motifs that does not involve adaptation. Divaricate architecture could be due to zones of inhomogeneity in the growing mantle that arise from interference patterns around regularly spaced centres. This architecture may sometimes turn out to have adaptive and functional value. Wavy ribs and other patterns might help creatures buried in the sand dig their way out (even though this may not be the most efficient means of doing so). *Pteria zebra* might use its shell's ramifications as camouflage when it disappears within an arborescent colony of Hydrozoa. Herring-bone mineralisation patterns might be of use to *Corculum cardissa* only. This shell lies flat on the seabed, its posterior surface, which bears the herring-bone pattern, facing upwards. Whereas the remainder of the shell is opaque, these patterns are translucid windows that provide light to the underlying algae that live in symbiosis with *Corculum.*

Seilacher's conclusion is clear. All morphological traits do not have to be interpreted in terms of use and adaptation. He suggests that three groups of factors might intervene in morphogenesis: historical phylogenetic factors, ecologic adaptive factors, structural (in German, *bautechnischer*) factors, which we shall refer to by the expression Seilacher's triangle.

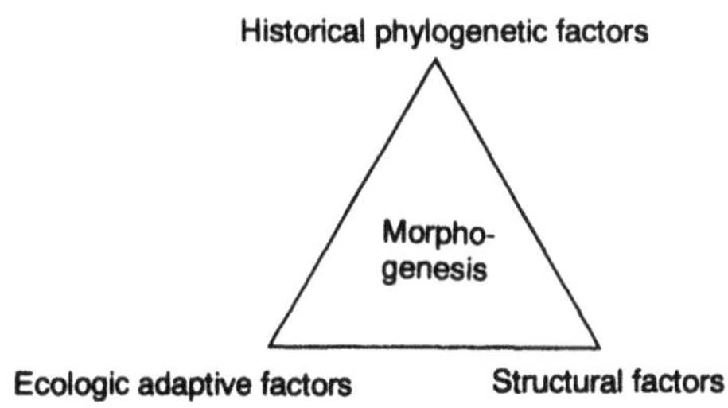


*Figure 7.3.* For the palaeontologist A. Seilacher, three groups of factors are involved in building the morphological traits of a species: historical and phylogenetic factors, ecologic and adaptive factors, and structural factors.

Let us consider each of these factors in turn:

1. Historical and phylogenetic factors. Palaeontologists do not describe fossils, their distinctive features and their arrangement in layers over time just for the purposes of stratigraphy. They also try and explain how shapes,

in particular body plans, come into being and develop. History (all the events that have taken place in a lineage) and phylogenesis (kinships) are key aspects of their research.

2. Ecologic and adaptive factors. For the most part, organisms appear well adapted to their environment "*as if each form were the consequence of external circumstances*".[6] The palaeobiological approach (also called ethological palaeontology) contributed extensively toward the success of palaeontology in the first half of the 20th century. A major limitation, however, was that it gave preference to elements of functional convergence and disregarded non functional observations:

> *Neither morphogenesis nor any function exists for its own sake. They are there to ensure, directly or indirectly, that an organism adapts to the outside world or, at least, maintains itself there.*[7]

Fortunately, since the 1950s, the concept of functional morphology (*Funktionsmorphologie*) is used in a wider context.

3. Structural aspects. The structural approach is a compromise between looking at body plans, on the one hand, and the relations between environment and adaptation, on the other. Organisms undergoing evolutionary change are governed by the structures permitted by morphology and morphogenesis (*bautechnische Lizenzen*). Adaptation is not omnipotent; neither is the body plan a choker.

A. Seilacher grouped phylogenetic, adaptive and architectural features together under the general term *Konstruktionsmorphologie*. An organism's morphological position within Seilacher's triangle is the result of a compromise between these three factors.

## 3. CONSTRAINTS AND BIOLOGICAL EVOLUTION

The notion of developmental constraint, nowadays widely accepted by evolutionists, has links with Seilacher's *Konstructionsmorphologie*. Inspired by Seilacher's three classes, S. J. Gould has proposed three ways of broaching the evolution of living things:[8]

1. Functionalism is the conventional view of adaptation found in the natural theology of the Anglican Archdeacon William Paley (1802) and in Lamarck's and Darwin's work.

2. Historicism is restricted to the study of descendancy and kinship among living things (J. Maynard Smith calls these local constraints[9]);

3. Formalism considers the properties of organisms as physical consequences of their material structure (J. Maynard Smith speaks of universal constraints).

The last two approaches, of which neo-Darwinists are rather dismissive, are particularly important in developmental studies. They help explain both the production of a variety of phenotypes and the limited diversity of phenotypes within a body plan, and they are directly related to the notion of developmental constraint. The question we must ask is, when a species' morphology undergoes no change over a very long time-span, even millions of years, is the stable state (or stasis) due to developmental constraints that limit morphological change or to selection that stabilises and maintains uniformity? How can we differentiate between developmental and selective constraints?

J. Maynard Smith and his colleagues suggested two ways of tackling the issue. The first approach was to compare predicted adaptation with reality (adaptation as recorded *a posteriori).* It can be equally well applied to the study of a population's reproduction mode as to the emergence of functional structures like those used in locomotion. The underlying question is the same. Does the problem facing the species or population determine the direction evolution will take? J. Maynard-Smith's second approach was to examine whether a single trait is encountered in independent branches of a phylum as a response to the same environmental need. This second approach established that selection, rather than developmental constraints, determine the predominance of certain traits. J. Maynard-Smith concluded that developmental constraints play a minor role in evolution compared to the dominant role of selection acting on genes, cells, *etc.* From then on, they were seen in a rather negative light; their role was to limit, even impede, the emergence of certain phenotypes to which selection might give rise.

S. J. Gould does not agree. In his mind, non selective constraints have a positive role. They make up the space where changes in living beings can emerge, develop or be perpetuated. They are not secondary to selection. This view is supported by a study of the *Cerion,* a species of snail common in Western Asia. Constraints in shape, such as the twirl of the shell around the axis, as well as historical constraints, account for the natural variety encountered in the shells of one and the same population, among geographically distinct exemplars of a single species, and among multitaxonomic fauna. The end of S.J. Gould's article is worth citing:

> *Nikolai Vavilov (1922) advanced too grandiose vision when he hoped that a specification of genetic and developmental channels might establish a 'periodic table' of possible forms, thus making biology as predictive as the chemistry of his countryman Mendeleev. But we may rightly hope for a predictive geometry of consequences when we can discern the formal and phyletic rules of development.*[10]

The debate is far from closed and has, recently, been fuelled by two fields of investigation: body plans and coevolution.

## 4. WAS LIFE WONDERFUL IN THE CAMBRIAN?

The introduction and conclusion of S. J. Gould's book "*Wonderful Life*" strangely echo J. Monod conclusions in "*Chance and Necessity*".

> *The old alliance is broken. At last, Man knows that he stands alone amidst the indifferent immensity of the Universe from which he emerged by chance. Neither his destiny nor his duty are inscribed anywhere. It is for him to choose between the Kingdom and darkness.* (J. Monod)[11]

> *Wind back the tape of life to the early days of the Burgess Shale; let it play again from an identical starting point, and the chance becomes vanishingly small that anything like human intelligence would grace the replay.* (S.J. Gould)[12]

> *The survival of Pikaia* [an ancestor of vertebrates] *was a contingency of "just history". I do not think that any "higher" answer can be given, and I cannot imagine that any resolution could be more fascinating. We are the offspring of history, and must establish our own paths in the most diverse and interesting of conceivable universes - one indifferent to our suffering, and therefore offering us maximal freedom to thrive, or to fail, in our own chosen way.* (S.J. Gould)[13]

We should examine carefully how the attitudes of the American palaeontologist and French biologist differ. Gould adopts a historical attitude and employs the notion of contingency. He distinguishes contingency from chance ("the instantaneous throw of the dice") which takes no account of initial conditions. In brief,

> *a historical explanation does not rest on direct deduction from laws of nature, but on an unpredictable sequence of antecedent states, where any major change in any step of the sequence would have altered the final result. The final result is therefore dependent, or contingent, upon everything that came before - the unerasable or determining signature of history.*[14]

These thoughts are echoed by Jean-Paul Sartre who, after contemplating the root of a chestnut tree, defines contingency:

> *A circle is not absurd and can be explained very well by rotating the segment of a line around one of its ends. Yet, a circle does not exist. On*

*the other hand, this root existed in that I could not explain it. Knotty, inert, nameless, it fascinated me, filled my eyes, and kept on reminding me of its existence. Although I repeated: "it is a root", this was of no avail. I saw clearly that one could not go from its function of a root or suction pump to this, this hard, compact seal-skin, this oily, callous and stubborn aspect. The function explained nothing. It helped explain what a root was but did not explain this particular root. This root with its colour, shape, petrified stance was ... beyond explanation.*[15]

*Contingency is essential. I mean that, by definition, existence is not a necessity. To exist is simply to be there. Those who exist appear, are encountered, but can never be deduced.*[16]

Going back to our fossils, let's take a brief look at the famous Burgess Shale upon which S.J. Gould's demonstration is based. The Burgess Shale is a fascinating story in the recent history of research in palaeontology. At the end of the autumn of 1909, the director of the Smithsonian Institution, Charles Doolittle Walcott, went off to look for fossil-rich rocky formations in British Columbia (Canada). He discovered a bed of fossils with astonishing forms and tried to "shoehorn" them (to quote Gould) into the standard classification groups. According to Walcott, these marine invertebrates from the middle Cambrian period (some 530 million years ago) were primitive representatives of the original animal groups which had gradually changed and diversified, that is evolved, to become current fauna. This perfectly orthodox view was not contested for more than fifty years.

In 1971, Harry Whittington published an essay that threw new light on the Burgess fossils. This was followed by works by Simon Conway Morris and Derek Briggs.[17] All these authors concluded unequivocally (some of their conclusions are contested today) that most Burgess organisms do not belong to any known zoological group. Their morphology is unrelated to that of any of the range of invertebrate forms currently inhabiting the oceans. Their body plans are astonishingly varied, for example,

*Opabinia, with its five eyes and frontal "nozzle"; Anomalocaris, the largest animal of its time, a fearsome predator with a circular jaw; Hallucigenia, with an anatomy to match its name.*[18]

In fact, the Burgess site not only displays spectacular palaeoecological aspects and exceptional conservation conditions but, most important of all, an extraordinary disparity of fauna. S.J. Gould suggests that this term should be used to describe the differences in anatomic organisation and the range of body plans.

The conclusions from studies of the Burgess fossils have been supported by excavation work in Greenland and China. Clearly, living things during the Cambrian period were much more dissimilar than we ever thought. One speaks of the "*zoological big bang in the Cambrian*" and of the "*Cambrian explosion*". There has been no slow, gradual diversification. On the contrary, nearly all the great body plans we know today (annelids, molluscs, echinoderms, chordates *etc.*), and many others which are now extinct, appeared in just a few tens of millions of years or less. The Burgess Shale contains at least twenty types of arthropods, each of them unique in its kind, in addition to representatives of the four big groups making up the branch.[19] The number of species within a phylum (which S.J. Gould calls diversity), however, is not very large:

> *The story of the last 500 million years has featured restriction followed by proliferation within a few stereotyped designs, not general expansion of range and increase in complexity as our favored iconography, the cone of increasing diversity, implies.*[20]

S.J. Gould calls this restriction decimation and illustrates it as shown in the diagram.

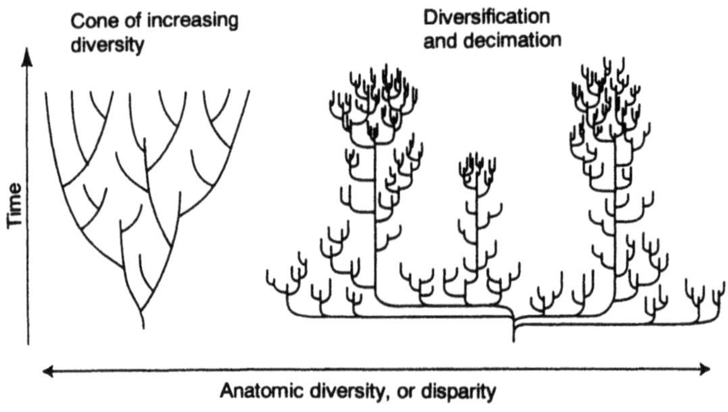


*Figure 7.4.* On the left, the false but commonly encountered iconography of the cone of increasing diversity. On the right, the revised model of diversification and decimation based on the proper reconstruction of the Burgess Shale. (Adapted from Gould, S.J., 1990, Wonderful Life, p. 46[12]).

Once we accept the idea of decimation, we can no longer interpret evolution in the light of a high level of adaptation. All body plans and all ensuing living forms had the same chances of survival. The number of plans far exceeded the current number and the "space" of the phylogenetically possible was wide open. Were the body plans that are now extinct decimated by pure chance, because of a kind of pitiless lottery? It is hard to say. However, because decimation was so extensive, we must consider all the body plans that might have been possible; they were the "initial conditions" from which history made its contingent selection.

We can now pinpoint the difference in Monod and Gould's views. Current living things appeared not so much because of chance but because of a strong historical contingency. In other words, if forms were randomly pulled out of a bag and then sorted by selection, this occurred within a space of possibilities limited by the constraints inherent to body plans. This space was brutally reduced at the end of the Cambrian period when many plans went extinct. Other worlds could have existed because the decimated forms did not disappear as a result of lesser adaptation. The idea of adaptation itself, so perfect, does not fit in with what actually happened. Evolution is more like a running race against environmental changes which, at times, can cause decimations and mass extinctions [21] or can be viewed as "*a history of organisms finding devious routes for getting round constraints*".[22]

## 5. STRUGGLE FOR LIFE AND A ROYAL RACE

That which applies to relations between a species' evolutionary fate and environmental changes can be extended to apply to relations between the evolutionary fates of interacting species. These two kinds of relations do they not belong to the more general concept of a trophic and biocenotic relationship? Charles Darwin had already raised the idea in *The Origin of Species*:

> *Thus I can understand how a flower and a bee might slowly become, either simultaneously or one after the other, modified and adapted to each other in the most perfect manner, by the continued preservation of all the individuals which presented slight deviations of structure mutually favourable to each other.*

Darwin spoke of coadaptation, his successors will speak of coevolution, a term coined by Leigh Van Valen, a researcher at Chicago University. In 1973, Van Valen determined the survival curves of about 50 groups of current or extinct living things (protists, plants and animals). He noted that, however long they had been alive, the likelihood of extinction was the same for each group. To explain this law of constant extinction, he proposed the Red Queen hypothesis. The "effective environment" of a homogeneous group of organisms changes according to a stochastic rate which is, on average, constant. Efforts to adapt must therefore be renewed constantly and the risk of extinction is always the same. "Effective environment" referred chiefly to other living organisms.[23] Changes in the physical environment could be disregarded because biotic factors on their own can generate, at this level, a continuous, self-steered movement in environment and species.

To understand the hypothesis, let's recall the story after which it is named. Led by the Red Queen, Alice runs until she is out of breath in order to stay in the same spot (illustrated by John Tenniel). We shall then show how it can help explain coevolution.

**Alice and the Red Queen**

> *"...Just at this moment, somehow or other, they began to run. Alice never could quite make out, in thinking it over afterwards, how it was that they began: all she remembers is, that they were running hand in hand, and the Queen went so fast that it was all she could do to keep up with her: and still the Queen kept crying, "Faster! Faster!" but Alice felt she could not go faster, though she had no breath left to say so.*
>
> *The most curious part of the thing was, that the trees and the other things round them never changed their places at all: however fast they went, they never seemed to pass anything. "I wonder if all the things move along with us?" thought poor puzzled Alice. And the Queen seemed to guess her thoughts, for she cried, "Faster! Don't try to talk!"*
>
> *Not that Alice had any idea of doing that. She felt as if she would never be able to talk again, she was getting so out of breath: and still the Queen cried, "Faster! Faster!" and dragged her along. "Are we nearly there?" Alice managed to pant out at last.*
>
> *"Nearly there!" the Queen repeated. "Why, we passed it ten minutes ago! Faster!" And then ran on for a time in silence, with the wind whistling in Alice's ears, and almost blowing her hair off her head, she fancied.*
>
> *"Now! Now!" cried the Queen. "Faster! Faster!" And they went so fast that at last they seemed to skim through the air, hardly touching the ground with their feet, till suddenly, just as Alice was getting quite exhausted, they stopped, and she found herself sitting on the ground breathless and giddy.*
>
> *The Queen propped her up against a tree, and said kindly, "You must rest a little now."*
>
> *Alice looked round her in surprise. "Why, I do believe we've been under this tree the whole time! Everything's just as it was!"*
>
> *"Of course it is," said the Queen: "what would you have it?"*
>
> *"Well, in our country," said Alice, still panting a little, "you'd generally get to somewhere else, if you ran very fast for a long time, as we've been doing."*

*"A slow sort of country!" said the Queen. "Now here, you see, it takes all the running you can do, to keep in the same place. If you want to get somewhere else, you must run at least twice as fast as that!"*

*"I'd rather not try, please!" said Alice. "I'm quite content to stay here - only I am so hot and thirsty!"* [24]

*Figure 7.5.* Alice and the Red Queen

Living species are the first environmental factor to be considered in an ecosystem. Influences from other species are not just due to prey-predator relationships. When a species evolves, it changes the effective environment, and this change then influences it in turn. Even if the physical environment were to remain stable, living things would still continue to evolve through mutual influence and adaptation. This is called coevolution.

The Red Queen hypothesis is profoundly Darwinian. It confers autonomy to living things with respect to nonbiotic factors and, especially, lacks any notion of perfection through adaptation. The aim is not to improve how one fits in with natural conditions but to remain alive despite these conditions. There is no room for anticipation; adaptations that grant greater future safety cannot be anticipated. All species evolve constantly but this evolution is just *"running [all one can] to keep in the same place."*

The Red Queen hypothesis is difficult to test as J. Maynard Smith and N.C. Stenseth showed in 1984.[25] It is often a possible mechanism but not always a necessary mechanism. There are, for instance, ecosystems of two or more species that could evolve indefinitely (as some host-pathogen duos). There are also times when the evolutionary process wanes when each species has reached its selective local optimum. Myxomatosis kills fewer rabbits now than some decades ago because selection forces are weaker and the two protagonists evolve more slowly. Acknowledging that the law of constant extinction was valid, N.C. Stenseth and J. Maynard-Smith proposed an alternative between two models, the stationary model and the Red Queen model.

1. In the stationary model, evolution gradually grinds to a halt when there are no physical changes. This happens in test tubes when self-replicating RNA molecules undergo selective constraints; each species reaches its local optimum. Evolution during the last three million years would thus be due to physical changes in the environment.

2. In the Red Queen model, coevolution could go on indefinitely, even in the absence of physical changes.

Can we choose between these two models? J.N. Thompson and others have suggested that the different types of species interactions (competition, parasitism, predation, *etc.*) in the Red Queen model can be seen as coevolutionary processes. On the other hand, S.J. Gould and N. Eldredge suggest in their theory of punctuated equilibria based on substantial palaeontological data[26] that only a change in nonbiotic environmental conditions, for example a change in climate, could cause an observable evolutionary change.

## 6. FINALISM REVISITED

The number of living forms depends on the limit history imposes on body plan number. Diversity is not limitless but restricted and due to variations on a set of morphogenetic themes rather than in actual body plans. A few main themes are modified in a multiplicity of ways. We have just to look at the number of known insect species. David Raup, who teaches statistics in palaeontology at the University of Chicago, has shown that only three variables are needed to study shell whorls: degree of curvature, translation of the point describing the spiral around the axis, and the distance of this point from the axis. If we build the cube representing the space of morphological possibilities using these three variables, we notice that living and fossil shells occupy only a few zones within the cube.[27] How can we explain such confined zones when nature is so profuse and diverse? Is it because each of the three factors of Seilacher's triangle has as great an impact on the emergence of new forms? It is, affirms S.J. Gould who reminds us that evolutionists have provided two kinds of explanation for the emergence of life forms. The first belongs to the Panglossian and adaptationist paradigm (as in the synthetic theory of evolution). By answering "why" the forms exist, we shall discover why some zones are empty. The other explanations refer to "how" the observed forms came into being, how the options in use depend on history, architecture and environment. Is this a move to avoid the issue of finality in biology or to tackle it anew?

The development of scientific thinking has discredited finality and teleology, and advantaged the study of proximal causality. In the past, we spoke of order and finality in fields where we now only speak of statistical regularity or the results of chance. Did we not used to say, like Aristotle who claimed that a free-falling stone was trying to return "home", that living beings had tendencies and intentions, in other words all kinds of things that we are no longer allowed to say today? More than 300 years BCE, Aristotle developed his concept of final cause in *The Parts of Animals*. How can we deny that the eye is made for seeing! Two centuries later, Lucretius criticised this concept in a chapter entitled "Some vital functions" (*De Rerum Natura):*

*We crave that thou wilt passionately flee*
*The one offence, and anxiously wilt shun*
*The error of presuming the clear lights*
*Of eyes created were that we might see;*
*Or thighs and knees, aprop upon the feet,*
*Thuswise can bended be, that we might step*
*With goodly strides ahead; or forearms joined*
*Unto the sturdy uppers, or serving hands*
*On either side were given, that we might do*
*Life's own demands. All such interpretation*
*Is aft-for-fore with inverse reasoning,*
*Since naught is born in body so that we*
*May use the same, but birth engenders use:*
*No seeing ere the lights of eyes were born,*
*No speaking ere the tongue created was;*
*But origin of tongue came long before*
*Discourse of words, and ears created were*
*Much earlier than any sound was heard;*
*And all the members, so meseems, were there*
*Before they got their use: and therefore, they*
*Could not be gendered for the sake of use.*[28]

For Lucretius, the statements "eyes created that we might see" and "no speaking ere the tongue created was" are equivalent. He seems to ignore that whereas all eyes in the animal kingdom are for seeing, most of the tongues are not used for talking ... The organ is "providential" in the etymological meaning of the word. Nature supplies organs, animals use them. Bernardin de Saint Pierre, who was for a short time superintendent of the Jardin des Plantes (Paris), made a similar misinterpretation in reverse. He wrote *Harmonies of Nature* half a century after Voltaire wrote *Candide* and yet insisted that melons, which display such beautifully regular ribs, are for family meals and pumpkins for meals among friends; better still, nature cares for our "moral enjoyment": "*the laurel for victory, the olive for peace, the palm for glory*". Because observers of nature have no explanatory mechanism, they have no alternative but to choose between a final cause inherent to life that induces and shapes the organ, and providence which anticipates our needs by providing the organ.

Now that we have evolutionary theory, we can at last elaborate "a science of biological finality". It is mutation and selection combined that has given birth to current life forms and to the wonderful ways in which they adapt. Each mutation occurs at random; it is not directed toward any end. The direction is given later by selection. We can thus say that both Aristotle and Lucretius were right. At the microscopic level of mutation, Lucretius was right in claiming that the structures are always there before they come into use. At the macroscopic level, Aristotle was right because a complex organ is always geared toward a function.

Because teleology was opposed to mechanisms, it could be nothing but a doctrinaire belief in a hidden, immanent or transcendent goal. This meant that any mention of a final cause was outlawed, especially in the teaching of biology. An English physiologist said in the 1950's that finalism is like a woman of ill repute with whom we cannot do without but with whom it is forbidden to be seen in the street. We had to satisfy ourselves with the rather facile distinction between the "how" of science and the "why" of philosophy and religion.

Because finality implies anticipation, it is related to time. It is virtually impossible to perform biological observations without taking the passing of time (cyclic or linear) into account. We do not mean the time of physics but the irreversible time of existence and history. In the 19th century, Claude Bernard had already perceived that physiological mechanisms need to be studied as if they depended on the achievement of a goal. For him, this goal was perpetuation of the internal medium.

**Claude Bernard and finalism**

*Physiologists and physicians must never forget that a living being is an organism with its own individuality. Since physicists and chemists cannot take their stand outside the universe, they study bodies and phenomena in themselves and separately without necessarily having to connect them with nature as a whole. But physiologists, finding themselves, on the contrary, outside the animal organism which they see as a whole, must take account of the harmony of this whole, even while trying to get inside, so as to understand the mechanism of its every part. The result is that physicists and chemists can reject all idea of final causes for the facts* ***that they observe; while physiologists are inclined to acknowledge an harmonious and pre-established unity in an organized body, all***

> ***of whose partial actions are interdependent and mutually generative.*** *We really must learn, then, that if we break up a living organism by isolating its different parts, it is only for the sake of ease in experimental analysis, and by no means in order to conceive them separately. Indeed when we wish to ascribe to a physiological quality its value and true significance, we must always refer it to this whole, and draw our final conclusion only in relation to its effects in the whole. It is doubtless because he felt this necessary interdependence among all parts of an organism, that Cuvier said that experimentation was not applicable to living beings, since it separated organized parts which should remain united.*[29]

By having a linear conception of time, we give it a sign or direction and call the changes it permits either improvement, progress or regression. Let us consider, for instance, the passage of autotrophous cells, which can synthesise all they require, to heterotrophous cells and organisms, which rely on organic molecules produced by autotrophous life forms. People who "*consider that autotrophism is vastly superior because it confers autonomy on organisms*"[30] do they not run the risk of attributing time a negative role and of seeing life's inventions as just ways of compensating weakening natural aptitudes? The direction given to time is usually positive, even if progress is relative, as the naturalist Ernst Mayr (a founder of the synthetic theory of evolution) reminds us:

*Evolutionary progress in one character is often paid for by weakening in another character. Obviously, general, global progress does not exist.*[31]

Introducing time into the natural sciences led to many attempts to invent laws that would explain the sequences in the history of life and, even, predict their outcome. We have the laws of increasing size (Deperet's law), increasing complexity and brain development (Teilhard de Chardin), recapitulation (Haeckel's law), and irreversibility (Dollo's law). Most were drawn up with a determinist and finalist outlook, and must be put into perspective. However, rather paradoxically, E. Mayr has stated that:

*Evolutionary biology, more than any other science, can provide valid answers to questions of the type: What is the object of the universe? What is the object of man? Why am I alive?*[32]

Is this the sign of the pride of a biologist who believes that the synthetic theory of evolution can suggest some answers to many questions that philosophers have been asking since pre-Socratic times whereas the question why, in the sense with what objective, has no meaning in the world of inanimate objects[33] (a world which, for example, physicists and astronomers study). E. Mayr denied adhering to any kind of finalism. In particular, he underscored that:

> *founding the analysis of a behaviour on terminology from human psychology inevitably leads to difficulties [...] Similarly, the assumption that there is an intention or a goal in every directional process encourages a teleological interpretation of organic evolution. This interpretation cannot be confirmed.*[34]

We shall not dwell on the contradictions in these two statements by Mayr but focus on the four life process categories, whether really or only seemingly directed, which he distinguishes:

- teleomatic processes, physicochemical processes which lead to a starting-point that is the consequence of a natural law, for example the law of gravity;
- teleonomic processes whose direction depends on a work-programme, that is coded or preordained information which control a phenomenon or behaviour;
- adapted or functional systems which are the fruits of the natural selection process and which are not always as perfect as we claim them to be;
- evolutionary progression, with the key question that there might be finality to the world, to which Mayr replies in the negative.

**The rendez-vous between fossils and computation**

*To test is to verify. To verify is to establish facts which we have produced or predicted. We can produce facts in physiology as in physics and in chemistry. In paleontology we predict facts not by anticipating the future but by rediscovering the past. If an experiment is defined as the testing of a hypothesis, we can say that, in some of its aspects, paleontology is an experimental science. The paleontologist who predicts that, by applying the principle of homology to current species, he will discover their ancestral structures at a given time in the Earth's history is in a similar position to that of the astronomer who predicts, by calculation, the existence of a planet which he will then be able to discover with a telescope. Thus, to quote P. de Saint-Seine, fossils rendez-vous with computation.*

*The constancy of the links, through variations in form and function, display a continuity which is compatible with the emergence of a new structural arrangement. The ability to predict or, if one prefers, to test hypotheses suggests that there is an order in nature that enables us to find the same regularity in the history of life as in the stars.*[35]

Contemporary science has evolved in a way that, in a fashion, rehabilitates finality, even if we have to play with the meanings of the words teleology and teleonomy. The term teleonomy, introduced by Anne Roe, George Gaylord Simpson and Konrad Lorenz, owes its consecration to the success of Jacques Monod's essay *Chance and Necessity*. For Monod, teleonomy is one of the three general properties that characterise living beings and distinguish them from the rest of the universe. The other two are autonomous morphogenesis and reproductive invariance. Teleonomy assumes the subjective idea that there is a plan:

> *We arbitrarily choose to define the essential teleonomic plan as the transmission from one generation to the next of the invariance content that characterises the species. All structures, all performances, and all activities that contribute toward the success of the essential plan are said to be "teleonomic".*[36]

Monod's main thrust is to link invariance and teleonomy, whilst having in mind the "*preservation and multiplication of the species*".[37] He can thus criticise afresh the dilemma between vitalism and animism, between intentionality within life and the intervention of an external intelligence. He concludes his book as follows. Having emerged by chance, man is free to choose his own plan because there is no other, neither a vital nor a divine plan.[38]

Science must rely on the principle of objectivity and cannot advance using final causes. When we investigate life, we cannot avoid the notions of finality and teleology because there is a finality (or rather several finalities) *de facto* but under no circumstance, does this imply that there is a conscience. The fundamental invariance of DNA, tied to selection, comes before everything else in teleology or teleonomy ("*only chance is the source of novelty or creation in the biosphere*").[39]

Monod's natural philosophy deserves attention and respect. However, it overemphasises the duo chance/necessity and disregards initial conditions and chain reactions. Of course, Monod was interested in the biochemical processes involved, in the "how" of things, but maybe, like adaptationists, he was too quick off the mark to answer the question "why".

One cannot be exhaustive on the subject of finality. We have noted, however, that the refusal to use this term or function, as the French still often do, in biology is incongruous. We have also noted that, as Gould, Lewontin or Seilacher urge us to do, we should look at finality in the (at least triangular) context of several interlinked constraints and causes acting at different levels. Adaptation is just one of them. We shall conclude by saying that evolutionary genetics must ally a fundamental interest in genetic information to the consideration of constraints.

## NOTES


1. Gould, S.J., and Lewontin, R.C., 1979, The spandrels of San Marco and the Panglossian paradigm: A critique of the adaptationist programme, *Proc. Roy. Soc. Lond. B* 205: 581-598.
2. Gould, S.J., 1993, Fulfilling the spandrels of world and mind, in: *Understanding Scientific Prose* (J. Selzer, ed.), University of Wisconsin Press, Madison, p. 310.
3. Voltaire, 1930 (originally published in 1759), *Candide ou l'Optimisme. Contes et Romans* (Vol. 2). Series: Les textes Français, Fernand Roches, Paris pp. 10-13. The emphasis is ours.
4. Schopenhauer, A., 1906 (originally published in German in 1819), *The World as Will and Idea (Die Welt as Wille und Vorstellung),* translated by R.B. Haldane and J. Kemp (3 vols), London; Schopenhauer, A., 1907 (originally published in German in 1836), *On the Will in Nature (Über der Willen in der Natur),* translated by K. Hillebrand, London.
5. Gould, S.J., 1989, A developmental constraint in *Cerion*, with comments on the definition and interpretation of constraint in evolution, *Evolution* 43: 516-539; Seilacher, A., 1970, Arbeitskonzept zur Konstrucktionsmorphologie, *Lethaia* 3: 393-396.
6. Daqué, E., 1921, quoted by Seilacher, 1970, *ibid.*, p. 393.
7. *Ibidem*
8. Gould, S.J., 1989, *ibid.*, p. 517.
9. Maynard-Smith, J., Burian, R., Kauffman, S., Alberch, P., Campbell, J., Goodwin, B., Lande, R., Raup, D., and Wolpert, L., 1985, Developmental constraints and evolution, *Quat. Rev. Biol.* 60: 265-287 (p. 267).
10. Gould, S.J., 1989, *ibid.*, p. 537.
11. Monod, J., 1971, *Chance and Necessity*, Knopf, New York. Vintage Paperback, 1972. [Originally published in France as *Le Hasard et la Nécessité*, Edition du Seuil, Paris, 1970].
12. Gould, S.J., 1990, *Wonderful Life: The Burgess Shale and the Nature of History*, Hutchinson Radius, Great Britain (Penguin Books, 1991, p. 14).
13. Gould, S.J., *ibid.*, p. 323.
14. Gould, S.J., *ibid.*, p. 283.
15. Sartre, J.P., 1938, *La Nausée*, Series: Folio, Gallimard, Paris, pp. 184-185.
16. Sartre, J.P., *ibid.*, p. 187.
17. Conway Morris, S., and Whittington, H.B., 1979, The animals of the Burgess Shale. *Scientific American*, 240: 122-133.
18. Gould, S.J., *ibid.*, p. 14.
19. *Trilobita* (which went extinct at the end of the primary era), *Chelicerata* (spiders, scorpions and some other groups of fossils), *Crustacea* (crabs, lobsters, shrimps, barnacles) and *Uniramia* (insects, centipedes, millipedes).
20. Gould, S.J., *ibid.*, p. 208.
21. Gould, S.J., *ibid.*
22. Maynard-Smith, J., Burian, R., Kauffman, S. et al., 1985, Developmental constraints and evolution, *Quat. Rev. Biol.*, 60, 265-287 (p. 282). Phrase used to describe the opinion of Lewontin and of other biologists.
23. Van Valen, L., 1973, A new evolutionary law, *Evol. Theory* 1: 1-30.
24. Carroll, L., 1871, *Through the Looking-Glass and What Alice Saw There*, Macmillan, London.
25. Stenseth, N.C., and Maynard-Smith, J., 1984, Coevolution in ecosystems: Red Queen evolution or stasis? *Evolution* 38: 870-880; Hoffman, A., 1991, Testing the Red Queen hypothesis, *J. Evol. Biol.* 4: 1-7.
26. Eldredge, N., and Gould, S.J., 1972, Punctuated equilibria: An alternative to phyletic gradualism. in: *Models in Paleobiology* (T.J.M. Shopf, ed.), Freeman, Cooper & Co., San Francisco, pp. 82-115.

27. Gould, S.J., 1980, The evolutionary biology of constraint, *Daedalus* 109, 39-52.
28. Titus Lucretius Carus, *Of the Nature of Things*. Book IV. Translated by William Ellery Leonard, 1997. Gutenberg Etext.
29. Bernard, C., 1966 (originally published in 1865), *Introduction à l'Etude de la Médecine Expérimentale*, Belfond, Paris, pp. 154-155. (translated by H.C. Green, 1957, An *Introduction to the Study of Experimental Medicine*. Dover Publications, NY, pp. 88-89.). We have introduced the bold font.
30. Roger, J., 1976, *Paléontologie Evolutive*, Series: Biologie évolutive n°2, Masson, Paris, p. 91.
31. Mayr, E., 1981, *La Biologie de l'Evolution*, Hermann, Paris, p. 170. (Section: "Structure Conceptuelle": conferences given by Ernst Mayr at the Collège de France in 1975).
32. Mayr, E., *ibid.*, p. 89.
33. Mayr, E., *ibid.*, p.108.
34. Mayr, E., *ibid.*, pp. 112-113.
35. Piveteau, J., 1983, *Origine et Destinée de l'Homme*, Masson, Paris, p. XII.
36. Monod, J., *ibid.*, p. 30.
37. Monod, J., *ibid.*, p. 26.
38. This conclusion provoked many reactions not only from "vitalists" and "animists" but also from researchers. Ilya Prigogine and Isabelle Stengers, by entitling their book *The Common Alliance* leave no room for doubt (Prigogine, I., and Stengers, I., 1979, *La Nouvelle Alliance. Métamorphose de la Science*. Series: Bibliothèques des Sciences Humaines, Gallimard, Paris).
39. Monod, J., *ibid.*, p. 127.

# CONCLUSION

## FROM AN IDEAL TO A TINKERED WORLD

Times have changed since the publication of *The Origin of Species*. The idea of evolution, its theories and paradigm have continued to evolve during the century and a half that separates us from the Darwinian "scandal". However, wherever we turn, the debates are still as fierce now as they were then, both in the scientific community and in Western society. The conflicts between fixists and transformists, biometricians and geneticists have given way to discussions on neutralist theory and the molecular clock, selection levels and selfish replicators. Social Darwinism and eugenist currents seem to have regressed because of the opprobrium their excesses of the first half of the 20th century met with and because of the development of genetics. On the other hand, sociobiology, the decoding of the human genome, and the heralded end to Darwinism regularly make the headlines in our news media. Who can deny that the mystery of life is still far from being solved, even if everyone claims that they henceforth hold the key to the mystery? How else can we explain the vitality, even expansion, of creationist thinking, and the imperious need to nominate ethics committees which decide whether genetic engineering experiments are warranted? Today as yesterday, biological evolution continues to arouse mankind's deepest anxieties, those about his origins and future.

Readers should therefore not merely reflect on the evolutionary theories presented in this book from a scientific point of view. They should also ask themselves where they stand with regard to the Darwinian paradigm, the Red Queen and Selfish Gene hypotheses. We are inviting them to go from the best of worlds - for instance the one imagined by the partisans of social Darwinism - to the world of possibilities - that of the Red Queen, constraints and short-term finality. This new direction may cost a few reassuring "-isms" and supposedly eternal "-ologies" (the "Darwinian revolution" will probably cost as much as the "Copernican revolution") but is this reason enough to refuse stepping through the looking glass?

To conclude, we shall retell the old story of Lot and his daughters who, in order to ensure offspring to their father and clan, had no qualms in transgressing one of the most basic laws of human, even animal, societies, the one forbidding incest. Their action is no doubt blameworthy but it illustrates one of the central ideas of contemporary evolutionary genetics. If genetic information has to be passed on through generations, then living beings are ready to make many sacrifices and even contravene a universal moral law. Other examples abound in ancient literature especially in cultures that, by tradition, are less interested in the individual than in the survival and development of the tribe. A people or tribe is not just an assembly of physical bodies but also includes all the kinds of information that make up a society (arts, customs, wisdom, *etc.*). Such a view of the individual and society is increasingly rare, and this probably explains the embarrassment or revolt provoked by the idea of a selfish gene. These reactions should be set aside or transcended both in the name of scientific research and on the basis of a different understanding of the place occupied by a human being within the species. Would this new understanding call into question the many benefits that individualism has provided us with today? Certainly not. However, can we and should we continue thinking of the human being as a free and independent individual? Nowadays, there is no lack of calls for a renewed understanding of the relationships among individuals, societies and cultures and between mankind and other living beings.

The biologist's dream should not prevent him or her from the task that befalls every research worker and member of the human community, that of building that which is possible.

# MATHEMATICAL APPENDIX

## MA1. EVOLUTIONARY PRESSURES

### 1. Mutation

Let's take a gene *A* of frequency $p_g$ at generation *g* which mutates at a rate *u* at each generation. If the reversion rate (i.e. the number of *reverse mutations* which restore the initial function) is negligible, at the next generation the frequency of *A* will be:

$p_{g+1} = (1-u)\, p_g$

The mutation-induced change in the frequency of *A* from one generation to the next (the mutation pressure) equals:

$\Delta p = p_{g+1} - p_g = -up_g$

From the first relationship, we deduce that:

$p_g = (1-u)^g \times p_0$

To halve the frequency of A, a number of generations *g* is needed so that:

$p_g = p_0/2 = (1-u)^g \times p_0$

Thus:

$(1-u)^g = 1/2$ and $g \approx 0.7/u$

However, mutation does not always occur in just one direction. Reverse mutations do exist. For example, let's suppose that gene *A* mutates into allele *a* with a frequency *u*, and that *a* mutates into *A* with a frequency *v*. The mutation pressure then becomes:

$\Delta p = -up + v(1-p) = v - p(u+v)$

This formula cancels itself out when $p$ equals $v/(u+v)$. In theory, the system ends up as a stable equilibrium if we consider mutation forces only. However, in a natural population, the mutation pressure is so low that gene frequency cannot be kept at its equilibrium value. Other much stronger pressures will modify population structure.

## 2. Selection

The concept of the fitness of a genotype (denoted $w$) was introduced in order to study how the allele frequencies of each locus evolve under the influence of selection. Let's consider a population of haploid individuals with two alleles, A and $a$, with fitnesses $w_A$ and $w_a$ at a particular locus. At generation g, their frequencies are $p_g$ and $q_g$ respectively (so that $p_g + q_g = 1$).

The mean fitness of the population, i.e. the average number of descendants per individual, is denoted $w_g$.

$w_g = p_g \text{ x } w_A + q_g \text{ x } w_a$

In this case, the proportion of $A$ alleles at generation g + 1 is:

$p_{g+1} = p_g \text{ x } w_A / w_g$

We can then calculate the variation in the frequency of allele $A$ from one generation to the next under the influence of selection:

$$\begin{aligned}\Delta p = p_{g+1} - p_g &= p_g \text{ x } w_A / w_g - p_g \\ &= p_g / w_g \text{ x } (w_A - p_g \text{ x } w_A - q_g \text{ x } w_A) \\ &= p_g \text{ x } q_g / w_g \text{ x } (w_A - w_a)\end{aligned}$$

The sign of $\Delta p$ gives the direction of evolution. Since $w_g$ and the product $p_g q_g$ are always positive, $\Delta p$ has the same sign as the difference between the two fitnesses $w_A$ and $w_a$.

## 3. Migration

The simplest model to highlight the role of migration is the "island" model. If $p_i$ is the frequency of an allele in the population under study (the "island") and $p_e$ the frequency of this allele in migrants (which are present in a proportion $m$ per generation), then at the next generation:

$p_i' = (1 - m) \text{ x } p_i + m\, p_e$

The difference between the allele frequencies of the two populations decreases:

$p_i' - p_e = (1 - m) \text{ x } (p_i - p_e)$

and the change in allele frequency due to migration is:

$\Delta p_i = p_i' - p_i = m \text{ x } (p_e - p_i)$

Equilibrium is reached when $p_i = p_e$.

# 4. Genetic drift

Genetic drift refers to the random fluctuation of allele frequencies in the absence of other selective pressures. The smaller the population, the more effective is genetic drift.

Let's consider a population of $N$ diploid individuals. Each of these individuals is the outcome of two gametes taken at random from those of the previous generation, i.e. $2N$ gametes. If the proportion of allele $A$ at generation 0 is denoted $p_0$, the number $X$ of gametes carrying $A$ amongst those giving rise to the next generation follows a binomial law with parameters $2N$ and $p_0$. Its expected value and variance are as follows:

$E(X) = 2N p_0$ and $Var(X) = 2N.\, p_0 \times (1 - p_0)$

The proportion $p_1$ of allele $A$, its expected value and variance at the next generation are calculated by:

$p_1 = X/2N \quad E(p_1) = 2N p_0/2N = p_0 \qquad Var(p_1) = p_0 \times (1 - p_0)/2N$

The new proportion $p_1$ is thus distributed around the expected value $p_0$ with a standard deviation equal to $\sqrt{[p_0 \times (1 - p_0)/2N)]}$.

For example, let's suppose the value of $p_0$ is 0.5. The standard deviation is then $\frac{1}{2} \times \sqrt{(1/2N)}$ which, for a population of 50 million individuals, is $5.10^{-5}$. (This is higher than the mutation rate which is about $10^{-6}$ per generation). For a population of just 50 individuals, the standard deviation would be $5.10^{-2}$. Genetic drift is more effective in the second case.

## MA2. INBREEDING COEFFICIENT

The inbreeding coefficient (denoted $f$) is the probability that two alleles in an individual are copies of one and the same gene within an ancestral population.

Let's consider a population of $N$ diploid individuals ($2N$ genes). Each individual produces an infinite number of gametes. A generation corresponds to $2N$ gametes sampled from the gamete pool of the previous generation. If we assume that the $2N$ genes of the individuals of the initial generation weren't identical, the inbreeding coefficient ($f$) at generation 0 is $f_0 = 0$. Let $f_g$ be the inbreeding coefficient of generation g. Two situations can arise:

• the two gametes making up an individual at generation $g$ each carry, at a given locus, a copy of the same gene from the previous generation (probability of the event: $1/2N$). There is a probability of 1 that the individual's two genes are identical.

• the two gametes carry copies of distinct genes from generation ($g$-$1$) (probability of the event: 1 - 1/2N). The probability that they will be identical at generation $g$ is therefore $f_{g\text{-}1}$ (inbreeding coefficient of the previous generation).

Thus,

$$f_g = (1/2N) + (1 - 1/2N)\, f_{g\text{-}1}$$

whence:

$$f_g = 1 - (1 - 1/2N)^g$$

## MA3. LETHAL MUTATIONS

Let $s$ be the selective disadvantage of the diploid individuals carrying the mutated form $a$ of the gene under study (one or two copies if the allele is dominant, a single copy if it is recessive). The so-called normal form is denoted $A$. If $s$ equals zero, there is no disadvantage. If $s$ equals 1, the mutated form is lethal.

Let $u$ be the mutation rate of gene $A$ into form $a$ per generation. By combining the $\Delta p$ of mutation and that of selection and assuming that they cancel each other out, we can calculate the frequency $q$ of allele $a$:

- if allele $a$ is recessive: $q = \sqrt{(u/s)}$
- if allele $a$ is dominant: $q = 2u/s$

## MA4. GENETIC LOAD

Genetic load (denoted $L$) is a decrease in the overall fitness of a population that is due to the different fitness levels of some individuals. It is given by:

$$L = 1 - w/w_{max}$$

where $w_{max}$ is the fitness of the best genotype(s) of the population and $w$ is average fitness. The load is thus the relative difference between the maximum fitness that a population of individuals with the best genotypes would have and average fitness.

Selection creates and maintains the load. Let's take the unlikely but simple case of extreme overdominance. The fitness of heterozygous ($Aa$) genotypes $w_2$ equals 1; those of homozygous ($AA$ and $aa$) genotypes, $w_1$ and $w_3$, are 0. Since only heterozygotes reproduce, they are at the origin of all progeny; polymorphism is thus maintained at $p_e = q_e = 1/2$. Overall, the zygotes produced at each generation are made up of 1/4 $AA$; 1/2 $Aa$; 1/4 $aa$.

At each generation, half of the population's individuals (the homozygotes) do not reproduce. This is the genetic load. In the case of sickle-cell anemia, the load is high and is the sum of the anaemic individuals *aa* and the homozygotes *AA* who have died from malaria.

Decreased fitness because of selectively maintained polymorphism at one locus can also occur at other loci. Individuals that are heterozygotes for one locus can be homozygotes for another locus, and can be wiped out because of the genetic load of this second locus. Clearly, under such circumstances, many loci cannot be maintained in a polymorphous state because too few members of the population would reproduce. To maintain 100 loci in a polymorphous state would require that the reproductive population falls to $(1/2)^{100}$, i.e. to about $10^{-30}$, members. This is, of course, impossible. To maintain just 10 loci polymorphous needs a decrease to $10^{-3}$. This is out of the question for many species including Man.

We have chosen the least favourable case. The differences in the fitness levels of homozygotes and heterozygotes can be very much smaller (but, nevertheless, not too small, as then there would be genetic drift). However, even then, selection with overdominance could not maintain polymorphism at several hundred loci.

## MA5. FIXATION OF NEUTRAL ALLELES

If, in a population of $N$ individuals ($2N$ genes), the rate of neutral mutations is $v$, then $2Nv$ new neutral genes will be produced at each generation. Each of the $2N$ genes, whether mutated or not, has the same probability ($1/2N$) of becoming fixed (this is what "neutral" means). A newly mutated gene will fix with a probability of $1/2N \times 2Nv = v$. This is true for each generation and thus, on average, a new neutral allele, which will replace the old one, will appear every $1/v$ generations regardless of population size. Using reaction-diffusion equations, Kimura has shown that, on average $4N$ generations will pass between the time a mutation appears and the time it becomes fixed. This is known as coalescence time. In most species, coalescence time is rather long and the fixed mutation is only temporary since it will be replaced by another neutral mutation some day. If the population is large enough ($4N>1/n$), the allele that will one day replace the one that is currently replacing an older allele has already appeared. Its frequency is on the increase. Consequently, there is polymorphism at all times with usually more than two alleles per locus.

# MA6. GAME THEORY

## 1. "Zero-sum" games

In "zero-sum" games, one player's gain is another player's loss. Participants' decisions are usually based on the *minimax* strategy. Because a player does not know what his opponent's next move will be, he judges the strength of his own moves by supposing that his opponent will play the best move. If player 1 employs strategy $i$ and player 2 strategy $j$, the gain of player 1 in the matrix of payoffs (which, by convention, always gives player 1's gain) is given by $a_{ij}$ (row $i$, column $j$). Player 2's gain is - $a_{ij}$.

In a *minimax* strategy, each player minimises the maximum loss his opponent can impose on him. Thus, player 1 searches for the lowest (minimum) result $Min_j(a_{ij})$ in each row $i$ (i.e. the greatest possible loss). He selects the row where this result is maximum $Max_i(Min_j(a_{ij}))$ (i.e. the least expected value of losing heavily). In contrast, player 2 must select $Min_j(Max_i(a_{ij}))$. If the two choices match, $Max_i(Min_j(a_{ij})) = Min_j(Max_i(a_{ij}))$.

Let's use the two following matrices as examples:

Game Matrix n° 1.

| | | | Player 2 | | |
|---|---|---|---|---|---|
| **Player 1** | column 1 | column 2 | column 3 | column 4 | $Min_j(a_{ij})$ |
| row 1 | +9 | +9 | -1 | +9 | (-1) |
| row 2 | +7 | +7 | **+7** | -2 | (-2) |
| row 3 | +2 | +2 | +2 | **+1** | (+1) |
| row 4 | +3 | -8 | -8 | -8 | (-8) |
| $Max_i(a_{ij})$ | (+9) | (+9) | (+7) | (+7) | |

Player 1 selects the third row and wins + 2. Player 2 selects the third column and loses 2.

Game Matrix n° 2

| | | | Player 2 | | |
|---|---|---|---|---|---|
| **Player 1** | column 1 | column 2 | column 3 | column 4 | $Min_j(a_{ij})$ |
| row 1 | +9 | +9 | +9 | -1 | (-1) |
| row 2 | +7 | +7 | -2 | -2 | (-2) |
| row 3 | +2 | +2 | +2 | **+1** | (+1) |
| row 4 | +3 | -8 | -8 | -8 | (-8) |
| $Max_i(a_{ij})$ | (+9) | (+9) | (+9) | (+1) | |

Player 1 selects the third row again and wins + 1. Player 2 chooses the fourth column and loses 1.

*The game is stable if any unilateral deviation from the strategy results in a loss* as in matrix n°2 where $Max_i(Min_j(a_{ij})) = Min_j(Max_i(a_{ij})) = +1$. It is not stable in matrix n°1 where these values are +1 and +7, respectively.

In matrix n°2, if the players play several times and only one of them changes strategy, he can but lose. Instead of winning 1, player 1 would lose 1, 2 or 8. Instead of losing 1, player 2 would always lose 2. (Let's not forget that player 1 defines his own strategy only, i.e. the rows, and player 2 can only choose among the columns). On the other hand, in matrix n°1, the situation is not stable. Aware that his opponent keeps on choosing column 3, player 1 might be tempted to play row 2. Instead of winning just 2, he would win 7. However, his opponent must not guess the change in strategy. If he does, he plays column 4 and player 1 loses 2 ... *The game is stable because neither player can change strategy without losing.*

## 2. Non zero-sum games

Mad driver game: It is a non-zero game because, as shown in the example below, when both players choose to go, they both lose (they lose 10).

| | | Alter | |
|---|---|---|---|
| | | Stop | Go |
| Ego | Stop | 0 | **-1** |
| | Go | **+1** | -10 |

Prisoner's dilemma is a non-zero game because, as illustrated by the example below, when one prisoner admits and the other denies, the former loses more (*i.e.* 20) than the latter wins (*i.e.* 1).

| | | Alter | |
|---|---|---|---|
| | | Admit | Deny |
| Ego | Admit | **-1** | +1 |
| | Deny | -20 | 0 |

All non-zero games with the following matrix are called prisoner's dilemma:

| | | Alter | |
|---|---|---|---|
| | | Betray | Cooperate |
| Ego | Betray | b | d |
| | Cooperate | a | c |

where $a<b<c<d$

## 3. Pure, mixed and conditional strategies

Pure strategies are encountered when an individual can have only one phenotype. Let's consider a population of individuals seeking to acquire a resource of unit value R (in terms of reproductive value) and where each individual can adopt either a H (Hawk) or D (Dove) strategy. H is an aggressive, risky strategy whereas d is a defensive, non aggressive strategy with no risk. When a hawk meets a hawk, it has a probability of 1/2 of winning resource R, and a probability 1/2 of losing it. If it loses, we assume it loses an amount P (in resource units). If it meets a dove, it wins the resource, at no cost. If a dove meets a dove, they share the resource (each gains R/2), and it costs them nothing. (NB: We are, of course, not describing interactions between real hawks and doves. The terms relate to the behaviour of individuals of the same species within a population.)

The matrix of payoffs is a prisoner's dilemma type matrix.

| | Hawk | Dove |
|---|---|---|
| Hawk | *(R-P)/2* | *R* |
| Dove | *0* | *R/2* |

The dove strategy is never stable even though it is the optimum strategy for the population,. In a population of doves, the hawk wins. The hawk strategy is stable only if *R* exceeds *P*. Thus, if the resource is greater than or equal to the risk of acquiring it by force, the aggressive strategy is the ESS. If, on the other hand, the resource is worth less than the possible acquisition cost, neither strategy is evolutionary stable. Consequently, doves are at an advantage in populations where there are many hawks, and hawks are at an advantage in populations where there are many doves. Polymorphism can thus be maintained under the action of frequency-dependent selection.

We have seen that when resource $R$ is less than amount $P$, the population cannot be made up of hawks only (everyone would lose out). We can calculate the equilibrium frequency $p_e$ of hawks; the average gain of hawks - $G(H)$ - then equals that of doves - $G(D)$:

$$G(H) = p \times G(H, H) + (1 - p) \times G(H, D) = p \times (R - P)/2 + (1 - p) \times R$$
$$G(D) = p \times G(D,H) + (1 - p) \times G(D,D) = p \times 0 + (1 - p) \times R/2$$
Equilibrium occurs when $p = p_e$, so that:
$$p_e \times (R - P)/2 + (1 - p_e) \times R = (1 - p_e) \times R/2$$
Thus: $p_e = R/P$

Let's suppose now that one individual may have several strategies. We need to find out how the individuals in the population will react. Let $p$ be the probability that they select the hawk strategy. The probability of selecting the dove strategy is then (1-p). Is there a value of $p$ that will ensure that the mixed strategy is an ESS? There is; a mixed strategy where hawk and dove strategies have probabilities of $R/P$ and $1 - R/P$, respectively, is an ESS. The optimal strategy (dove) is, of course, still not stable.

In nature, however, two "players" hardly ever meet by chance, and a player is most unlikely to use a probabilistic algorithm to make a choice. He will give due regard to the favourable conditions offered by the encounter and adopt a conditional strategy. J. Maynard-Smith has imagined a territorial (or bourgeois) strategy which involves behaving like a hawk when one thinks one was there first (the resident) and like a dove if one comes in second (the intruder). When two individuals taken at random meet, the probability that one arrived first is 1/2, regardless of its behaviour. We can thus calculate the gains to be expected under different circumstances. Let's suppose two bourgeois meet; they do not fight but behave like a hawk and dove, or a dove and a hawk. Thus, the mean gain of a bourgeois confronted by another bourgeois is $R \times (1/2) + 0 \times (1/2)$, i.e. $R/2$. If a bourgeois meets a dove, it behaves as if it were the owner - and thus like a hawk - one in every two times (gain = $R$), or it behaves like a dove (gain = $R/2$). Similarly, if a bourgeois meets a hawk, it acts like a hawk (gain = $(R - P)/2$) or a dove (gain = 0) one in every two times. Conversely, if a hawk meets a bourgeois, it is up against a hawk (gain = $(R - P)/2$) or dove (gain = $R$) one in every two times. Finally, if a dove meets a bourgeois, it is up against a hawk (gain = 0) or a dove (gain = $R/2$) one in every two times. The payoff matrix is thus as follows:

| | Hawk | Dove | Bourgeois |
|---|---|---|---|
| Hawk | *(R-P)/2* | *R* | *3R/4-P/4* |
| Dove | *0* | *R/2* | *R/4* |
| Bourgeois | *(R-P)/4* | *3R/4* | *R/2* |

If *R* is greater than *P*, the ESS is the hawk strategy. If *R* is less than *P*, the ESS strategy is the bourgeois strategy. The ESS strategy is better than all mixed strategies, in particular the one where *p* equals *R/P*.

# GLOSSARY

## A

**Alleles** (Bateson and Saunders, 1902): Homologous genes that ensure the same function in different ways. For example, the gene "pea colour" has two alleles: "yellow pea" allele/"green pea" allele.

**Allogamy** (also called outbreeding and sometimes cross-breeding): Mating of individuals with each other (cross-fertilization). The opposite of autogamy.

**Autogamy**: Ability of an individual to self-fertilise. The opposite of allogamy.

## C

**Character** or trait: Any attribute that can be observed in an individual, cell or molecule (size, colour, reaction to different factors, electrophoretic mobility).

**Chromosome**: A structure containing a single molecule of DNA (or RNA in some viruses) and which therefore bears genetic information. The eukaryote chromosome was defined initially as the cytological unit that, as a result of condensation, could be individualised during nuclear division (mitosis or meiosis). Chromosomes are distribution units during nuclear division.

**Codon**: A sequence of three nucleotides in DNA. It is an instruction to incorporate an amino acid into a polypeptide chain. There are 64 codons and only 20 amino acids. Several codons are therefore synonymous.

**Crossing-over**: Swapping of parts of homologous chromosomes by two non-sister chromatids of a bivalent. (A bivalent is a unit of paired homologous chromosomes during meiotic division.)

## D

**Deoxyribonucleic acid (DNA)**: A polymer of deoxyribonucleotides. Two complementary strands of DNA entwined in a double helix form the genome of most organisms.

**Diploid**: An individual with two copies of each chromosome in each cell.

**Dominance** (Mendel, 1866): Originally, the masking of the phenotypic effect of one allele (a recessive allele) by another (the dominant allele). By extension, the interaction between homologous genes. When the phenotype of heterozygote $a_1a_2$ is the sum of the phenotypes of homozygotes $a_1a_1$ and $a_2a_2$, alleles $a_1$ and $a_2$ are co-dominant. When it is intermediate between these phenotypes, they are semi-dominant.

## E

**Epistasis** (Bateson, 1907): Originally, the masking of the expression of a gene by another (epistatic) gene at a different locus. By extension, the interaction between non-homologous genes.

**Eukaryote**: An organism whose cell nucleus is separated from the cytoplasm by a membrane.

**Evolutionary pressure**: An action on a population that modifies allele structure (drift, selection, mutation, migration).

**Exon**: The protein-coding DNA sequence of a gene (i.e. the DNA sequence that is transcribed into messenger RNA (mRNA) which is translated into protein).

## F

**Fitness** (also called adaptive or selective value): The reproductive success of the individuals of a genotype submitted to selection. It is given by the average number of progeny at the following generation.

## G

**Gamete**: Reproductive cell with a haploid set of chromosomes.

**Gene** (Johannsen, 1909) from pangene (De Vries, 1903) from pangenesis (Darwin, 1868): Information carried by an ordered sequence of nucleotides in DNA, together with the signals required for their expression, enabling a basic function to be carried out (production of a polypeptide chain, of transfer RNA (tRNA) or ribosomal RNA (rRNA)). Homologous genes are genes which are normally paired up during meiosis and which have the same function. Genes that are identical through descent are copies, with no mutations, of the same ancestral gene.

**Gene coalescence** in a population (Kingman, 1982): All the homologous genes in a population of a limited number of individuals descend from a single ancestral gene which was present in the population's genes some generations earlier.

**Gene conversion**: Change in an allele sequence of a chromatid by DNA recombination. The sequence becomes identical to that of the chromatid with which pairing occurred during meiosis.

**Gene frequency**: For a population as a whole, the percentage of genes at a particular locus that are of one form (allele) as opposed to another.

**Genetic drift** (Wright, 1921): Random fluctuation in gene frequency over generations in a population of limited size. It is due to the sampling effect.

**Genetic load** (Muller, 1950): Decrease in the fitness of a population which is due to the maintenance of individuals with fitnesses that are lower than that of the individual in the population with the highest fitness.

**Genetic structure**: The allele structure and phenotypic structure of a population taken together.

**Genome** (Weinkler, 1920): All the genetic information borne by an individual. In a restricted sense, all the genes making up a functional unit (mitochondrial genome, genome inherited from an ancestral species in a polyploid individual, *etc.*)

**Genotype** (Johannsen, 1909): The allele composition of a locus (or loci) in an individual.

## H

**Haploid**: An individual with a single copy of each chromosome in each cell.

**Heterochromatin**: Highly condensed and inactive region of a chromosome.

**Heterogamy**: Mating of individuals of dissimilar phenotype.

**Heterozygote** (Bateson and Saunders, 1902): An individual which carries both members of a pair of alleles at a locus.

**Homogamy**: Mating of individuals of similar phenotype.

**Homozygote** (Bateson and Saunders, 1902): An individual which carries two genes of the same allele form at a locus.

## I

**Inbreeding**: Mating of genetically related individuals.

**Intron**: The DNA base sequence interrupting the protein coding sequences (exons) of a gene. Intron sequences are transcribed into RNA but are cut out of the message before it is translated into protein.

## K

**Kinship**: Related individuals, with common ancestors.

## L

**Linkage** (Morgan, 1911): Genetic link between two loci.

**Linkage disequilibrium** (Kimura, 1956): Non random association of alleles belonging to different loci in a population.

**Locus** (Morgan *et al.*, 1915): Originally, the position of a gene on a chromosome. The locus is defined as a homology class. It is named after the function of the constituent homologous genes. For example: the locus of pea colour.

## M

**Meiosis**: Two consecutive cell divisions leading to the production of gametes. It is characterised by the pairing up and segregation of homologous chromosomes in daughter cells. Each daughter cell takes half the genetic inheritance.

**Migration**: Genes passing from one population to another either as individuals or gametes.

**Mitosis**: Cell division which is accompanied by chromosome duplication and the distribution of the chromatids between daughter cells. All of the genetic information is passed on.

**Mutation** (De Vries, 1901): A sudden heritable change in genetic information. Mutations can affect nucleotide sequence, gene arrangement or the amount of genetic material.

## N

**Neutral theory**: A theory stating that much of the polymorphism observed in natural populations is selectively neutral and of no adaptive value.

**Nucleus**: An organelle harbouring the nuclear genome. During cell division, nuclear DNA is condensed into chromosomes and the nucleus disappears.

## O

**Organelle**: A differentiated cell constituent with a specific function (nucleus, mitochondria, chloroplasts).

**Overdominance** (Hull, 1946): A relationship in which, for a given character, the fitness of a heterozygote is higher than that of the two corresponding homozygotes.

## P

**Panmixis** (Weismann, 1895): Mating of gametes at random with regard to a locus (or loci).

**Phenotype** (Johanssen, 1909): The appearance of an organism with respect to a particular character or group of characters as a result of the interaction of its genotype and its environment.

**Pleiotropy** (Plate, 1910): The effect of a gene on several different characters.

**Polymorphism** (Ford, 1940): The coexistence of two or more discontinuous forms within a population in a ratio such that the rarer form cannot be maintained by recurrent mutation. Gene polymorphism is the coexistence within a population of several alleles at a locus.

**Polyploidy**: The presence of more than two copies of each chromosome in each cell. It often occurs in plants.

**Population** (Johanssen, 1903): A set of individuals that mate sexually with each other.

**Prokaryote**: A single cell organism with no well-defined nucleus.

# R

**Recessive, Recessivity** (Mendel, 1866): see Dominance.

**Recombination** (Bridges and Morgan, 1923): A process by which progeny derive new combinations of parental characters. The association of alleles of two linked loci found in the parents is modified and leads to "recombined" associations.

**Reproduction mode**: The way in which gametes unite, with regard to gene(s), to produce progeny.

**Restriction enzyme (site, map)**: A bacterial protein that recognises specific sequences of a few nucleotides and cleaves DNA at these sites. The positions of the sites on the chromosome provide a restriction map. Restriction enzymes probably evolved to protect bacteria from foreign DNA. Bacterial sites are not cleaved because they are specifically methylated.

**Ribonucleic acid (RNA)**: A polymer of ribonucleotides. RNA is the product of the transcription of DNA. The genome of some viruses is made up of RNA that can be retrotranscribed into DNA. Messenger RNA (mRNA) carries the information for one or more proteins. The passage from mRNA to protein is called translation.

# S

**Segregation** (Bateson and Saunders, 1902): Separation of allele forms that were initially combined at a locus.

**Selection** (Darwin, 1858): A process resulting in individuals surviving or reproducing better or worse than others, depending upon their genotype. Selection to improve species can be natural or artificial.

**Self-fertilisation**: Fertilisation between two gametes, one male, the other female, from the same individual.

**Sister chromatids**: Twin copies of a chromosome which remain closely associated when a chromosome duplicates. Non-sister chromatids belong to different chromosomes within a homologous pair.

**Strain**: A breed of individuals that are perpetuated over generations without any change in the genotype of the individuals. Pure strain: Breed of genetically identical, homozygous individuals maintained by inbreeding. Wild strain: Laboratory strain bred from individuals taken from nature and used as a reference.

## Z

**Zygote**: A diploid cell formed by union of a male gamete and a female gamete, which will produce an individual.

# BIBLIOGRAPHY


Atlan, Anne and Gouyon, Pierre-Henri, 1994, Les conflits intra-génomiques, *Société Française de Génétique*, pp.1–9.

Ayala, Francisco J., 1982, *Biologie Moléculaire et Evolution.* Series: Biologie Evolutive, N°7, Masson, Paris.

Barra, Jacqueline and Renard, Jean-Paul, 1986, Superflu, le spermatozoïde? *La Recherche* 17:264–266.

Belhassen, E. and Gouyon, P.-H., 1986, Flux géniques, perturbations, et gynodiœcie chez *Thymus vulgaris, Colloque national C.N.R.S. Biologie des populations, Sept. 1986*, pp.134–140.

Blanc, Marcel, 1984, Gregor Mendel: la légende du génie méconnu, *La Recherche* 15:46–59.

Blanc, Marcel, 1990, *Les Héritiers de Darwin. L'Evolution en Mutation*, Series: Science ouverte, Seuil, Paris.

Blunt, Wilfrid, 1986, *Linné, Le Prince des Botanistes*, Series: Un Savant, Une Epoque, Belin, Paris.

Brakefield, Paul M., 1987, Industrial melanism: Do we have the answers? *TREE* 2:117–122.

Buican, Denis, 1984, *Histoire de la Génétique et de l'Evolutionnisme en France*, Series: Histoires, Presses Universitaires de France, Paris.

Buican, Denis, 1987, *Génétique et Pensée Evolutionniste*, Series: Regards sur l'Histoire, SEDES, Paris.

Buican, Denis, 1988, *Lyssenko et le Lyssenkisme*, Series: Que sais-je? N° 2394, Presses Universitaires de France, Paris.

Burnet, Macfarlane, 1978, *Endurance of Life. The implications of Genetics for Human Life*, Cambridge University Press, Cambridge.

Carles, Jules, 1988, La génétique et les origines de l'homme», *Nouv. Rev. Théol.* 110:245–256.

Carles, Jules, 1994, *Le Premier Homme*, Series: Bref N° 49, Cerf/Fides, Paris.

Carton, Yves, 1988,La coévolution, *La Recherche* 19: 1022–1031.

Chandebois, Rosine, 1993, *Pour en Finir avec le Darwinisme. Une Nouvelle Logique du Vivant*, Editions Espaces 34 (Espace Science), Montpellier.

Coppens, Yves, 1983, *Le Singe, l'Afrique et l'Homme*, Series: Le Temps des Sciences, Fayard, Paris.

Couvet, Denis, Gouyon, Pierre-Henri, and Kjellberg, Finn, 1985, De la métapopulation au voisinage : la génétique des populations en déséquilibre, *Génét. Sél. Evol.* 17: 407–414.
Damuth, John, 1985, Selection among "species": A formulation in terms of natural functional units, *Evolution* 39: 1132–1146.
Darwin, Charles, 1859, *On the Origin of Species by Means of Natural Selection*, Murray, London.
Darwin, Charles, 1871, *The Descent of Man and Selection in Relation to Sex*, Appleton, New York.
Dawkins, Richard, 1986, *The Blind Watchmaker*, Longmans, London.
Dawkins, Richard, 1976 (revised 1989), *The Selfish Gene*, Oxford University Press, Oxford.
Dawkins, Richard, 1995, *River Out of Eden. A Darwinian View of Life*, Weidenfeld and Nicholson (Science Masters), London.
Denton, Michael, 1985, *Evolution. A Theory in Crisis*, Burnett, London.
Devillers, C. and Chaline, J., 1989, *La Théorie de l'Evolution. Etat de la Question à la Lumière des Connaissances Scientifiques Actuelles*, Dunod-Bordas, Paris.
Devillers, C. and Tintant, H., 1996, *Questions sur la Théorie de l'Evolution*, Presses Universitaires de France, Paris.
Doolittle, W. Ford and Sapienza, Carmen, 1980, Selfish genes, the phenotype paradigm and genome evolution, *Nature* 284: 601–603.
Eldredge, Niles and Gould, Stephen Jay, 1972, Punctuated equilibria: An alternative to phyletic gradualism, in: *Models in Paleobiology* (T.J.M. Schopf, ed.), Freeman, Cooper, San Francisco.
Fischer, Jean-Louis (ed.), 1991, *La Naissance de la Vie. Une Anthologie*, Presses Pocket (Agora-Les Classiques 12), Paris.
Gliddon, Chris J. and Gouyon, Pierre-Henri, 1989, The units of selection, *TREE* 4: 204–208.
Gould, Stephen Jay, 1977, *Ontogeny and Phylogeny*, Harvard University Press, Cambridge (Mass.).
Gould, Stephen Jay, 1980, The evolutionary biology of constraint, *Dædalus* 109: 39–52.
Gould, Stephen Jay, 1980, Is a new and general theory of evolution emerging?, *Paleobiology* 6: 119–130.
Gould, Stephen Jay, 1980, *The Panda's Thumb*, Norton, New York.
Gould, Stephen Jay, 1983, Unorthodoxies in the first formulation of natural selection, *Evolution* 37: 856–858.
Gould, Stephen Jay, 1977, *Ever since Darwin*, Norton, New York.
Gould, Stephen Jay, 1989, Developmental constraint in *Cerion*, with comments on the definition and interpretation of constraint in evolution, *Evolution* 43: 516–539.
Gould, Stephen Jay, 1990, *The Individual in Darwin's World*, Edinburgh University Press, Edinburgh.
Gould, Stephen Jay, 1989, *Wonderful Life. The Burgess Shale and the Nature of History*, Norton, New York.
Gould, Stephen Jay, 1995, *Adam's Navel and Other Books*, Penguin Books (Penguin 60s), London.
Gould, Stephen J., 1996, *Dinosaur in a Haystack. Reflections in Natural History*, Jonatham Cape, London.
Gould, Stephen Jay, Gilinsky, Norman L., and German, Rebecca Z., 1987, Asymmetry of lineages and the direction of evolutionary time, *Science* 236: 1437–1441.
Gould, Stephen Jay and Lewontin, Richard C., 1979, The spandrels of San Marco and the Panglossian paradigm: A critique of the adaptationist programme, *Proc. Roy. Soc. London* B205: 581–598.

Gouyon, Pierre-Henri, Gliddon, Chris J., and Couvet, Denis, 1988, The evolution of reproductive systems: A hierarchy of causes, in: *Plant Population Ecology* (A.J. Davy et al., eds), Blackwell Scientific Publications, London, pp. 23-33.
Gouyon, Pierre-Henri, Gliddon, Chris J., and Couvet, Denis,1988, Conflits nucléocytoplasmiques et stérilité mâle, Variabilité génétique cytoplasmique et stérilité mâle cytoplasmique, *Les Colloques de l'INRA* 45: 35–50.
Gouyon, Pierre-Henri, Maurice, Sandrine and Reboud, Xavier, 1993, Le sexe pour quoi faire? *La Recherche* 24: 70–76.
Gouyon, Pierre-Henri and Olivieri, Isabelle, 1989, Génétique évolutive et théorie des jeux, *Société Française de Biométrie* 8: 18–38.
Grassé, Pierre-Paul, 1973, *L'Evolution du Vivant. Matériaux pour une Nouvelle Théorie Transformiste*, Series: Sciences d'aujourd'hui, Albin Michel, Paris.
Gros, François, 1986, *Les Secrets du Gène*, Odile Jacob/Seuil, Paris.
Hoffman, Antoni, 1991, Testing the Red Queen Hypothesis, *J. Evol. Biol.* 4: 1–7.
Jacob, François, 1990 (1st ed. 1970), *La Logique du Vivant. Une Histoire de l'Hérédité*, Series: Tel. 2, Gallimard, Paris.
Jacob, François, 1981, *Le Jeu des Possibles. Essai sur la Diversité du Vivant*, Fayard, Paris.
Jaisson, Pierre, 1993, *La Fourmi et le Sociobiologiste*, Odile Jacob, Paris.
Kimura, Motoo, 1982, *The Neutral Theory of Molecular Evolution*, Cambridge University Press, Cambridge (Mass.).
Klein, Jan, Takahata, Naoyuki and Ayala, Francisco J., 1994, Reconnaissance de soi et origine de l'homme, *Pour la Science*, 196: 56–61.
Lamarck, Jean-Baptiste, 1994, *Philosophie Zoologique*, Flammarion, Paris.
Langaney, André, 1987 (1st ed 1979), *Le Sexe et l'Innovation*, Series: Point-Sciences, Seuil Paris.
Langaney, André, 1999, *La Philosophie ... Biologique*, Belin, Paris.
Langaney, André, Pellegrini, Béatrice, and Poloni, Estella, 1989, L'homme descend du sexe, *La Recherche* 20: 995–1007.
Lecourt, Dominique, 1992, *L'Amérique entre la Bible et Darwin*, Presses Universitaires de France, Paris.
Le Guyader, Hervé, 1998, *Geoffroy Saint-Hilaire, un naturaliste visionnaire*, Series: Un Savant, une Epoque, Belin, Paris.
Lemarchand, Fabienne, 1993, Explosion cambrienne: le rythme s'accélère, *La Recherche* 24: 1402–1403.
Lewin, Roger, 1993, Human Evolution: An Illustrated Introduction (3rd edition), Blackwell Scientific Publications, Oxford.
Maynard Smith, John, 1962 (1st ed.1958), *The Theory of Evolution*, Penguin Books, Baltimore (Mass.).
Maynard Smith, J., Burian, R., and Kauffman, S., 1985, Developmental constraints and evolution, *Quart. Rev. Biol.* 60: 265–287.
Maynard Smith, John, 1989, *Evolutionary Genetics*, Oxford University Press, Oxford.
Mayr, Ernst, 1970, *Populations, Species, and Evolution: An Abridgement of Animal Species and Evolution*, Harvard University Press, Cambridge (Mass.).
Mayr, Ernst, 1981, *Toward a New Philosophy of Biology: Observations of an Evolutionist*, Havard University Press, Cambridge (Mass.).
Mayr, Ernst, 1981, *La Biologie de l'Evolution*, Hermann, Paris.
Mayr, Ernst, 1985,Weismann and Evolution, *J. Hist. Biol.* 18: 295–329.
Medvedev, Zhores A., 1969 (transl.: Lerner, I. Michael), *The Rise and Fall of T. D. Lysenko*, Columbia University Press, New York/London.

Monod, Jacques, 1970, *Le Hasard et la Nécessité. Essai sur la Philosophie Naturelle de la Biologie Moderne*, Seuil, Paris. (Published in English, 1971, *Chance and Necessity*, Knopf, New York (Vintage paperback, 1972)).
Müller, Gerd B. and Wagner, Günter P., 1991, Novelty in evolution: Restructuring the concept, *Annu. Rev. Ecol. Syst.* 22: 229–256.
Nowak, Marin, May, Robert, and Sigmund, Karl, 1995, L'arithmétique de l'entraide, *Pour la Science*, 214: 56–61.
Olivier, Isabelle and Gouyon, Pierre-Henri, 1990, Evolution et théorie des jeux, *Science et Vie* 173: 144–146.
Orel, Vitezslav and Armogathe, J.R., 1985, *Mendel, un Inconnu Célèbre*, Belin, Paris.
Orgel, L. E. and Crick, F. H. C., 1980, Selfish DNA: The ultimate parasite, *Nature* 284: 604–607.
Pichot, André, 1995, *L'Eugénisme ou les Généticiens Saisis par la Philanthropie*, Series: Optiques Philosophie, N° 212, Hatier, Paris.
Piveteau, Jean, 1983 (1st ed. 1973), *Origine et Destinée de l'Homme*, Series: Abrégés, Masson, Paris.
Ridley, Mark, 1993, *The Red Queen: Sex and the Evolution of Human Nature*, Viking, London.
Robert, Jean-Michel, 1994, Le fardeau génétique, *APBG*, pp. 45–48.
de Rosnay, Joël, 1977 (1st ed 1966), *Les Origines de la Vie. De l'Atome à la Cellule*, Series: Points-Sciences, N° 10, Seuil, Paris.
Rostand, Jean, 1964 (1st ed.1945), *Esquisse d'une Histoire de la Biologie*, Series: NRF-Idées , Gallimard, Paris.
Ruse, Michael, 1982 (1st ed.1979), *Sociobiology, Sense or Nonsense?* D. Reidel (Pallas Paperback. 7), Dordrecht.
Sauclières, Gilles, 1988, Eve à la "une", *La Recherche* 19: 525–527.
Schrödinger, Erwin, 1944, *What is Life? The Physical Aspect of Living Cell*, Cambridge University Press, Cambridge.
Schützenberger, Marcel-Paul, 1996, Les failles du darwinisme, *La Recherche* 283: 87–90.
Seilacher, Adolf, 1970, Arbeitskonzept zur Konstruktionsmorphologie, *Lethaia*, 3: 393–396.
Serre, Jean-Louis,1984, La genèse de l'œuvre de Mendel, *La Recherche* 15: 1072–1080.
Tassy, Pascal and Darlu, Pierre, 1987, L'ADN, l'Afrique et l'homme, *La Recherche* 18: 979–981.
Tassy, Pascal, 1991, *L'Arbre à Remonter le Temps. Les Rencontres de la Systématique et de l'Evolution*, Series: Epistémè Essais, Christian Bourgois, Paris.
Tassy, Pascal, 1991, *Le Message des Fossiles*, Series: Questions de Science, Hachette, Paris.
Tastemain, Catherine, 1987, Les chromosomes se souviennent aussi, *La Recherche* 18: 1537–1539.
Taylor, Gordon Rattray, 1963, *The Science of Life. A Picture History of Biology*. Thames & Hudson, London.
Thuillier, Pierre, 1981, *Les Biologistes Vont-ils Prendre le Pouvoir? La Sociobiologie en Question*, Series: De la Science, Editions Complexe, Paris.
Tort, Patrick, 1983, *La Pensée Hiérarchique et l'Evolution. Les Complexes Discursifs*, Series: RES Résonances, Aubier Montaigne, Paris.
Tort, Patrick (ed.), 1985, *Misère de la Sociobiologie*, Presses Universitaires de France, Paris.
Tort, Patrick (ed.), 1991, *Darwinisme et Société*, Presses Universitaires de France, Paris.
Tort, Patrick, 1996, *Dictionnaire du Darwinisme et de l'Evolution*, Presses Universitaires de France, Paris.
Veuille, Michel, 1981, L'"adultère" chez les oiseaux: un concept trompeur, *La Recherche* 12: 880–881.

Veuille, Michel, 1986, *La sociobiologie*, Series: Que sais-je?, Presses Universitaires de France, Paris.
Vialleton, Louis, 1929, *L'Origine des Etres Vivants. L'Illusion Transformiste*, Plon, Paris.
Wallace, Alfred Russel, 1870, *Contributions to the Theory of Natural Selection: A Series of Essays*, C. Reinwald, Paris.
Wallace, Alfred Russel, 1912 (1st ed.1889), *Darwinism. An Exposition of the Theory of Natural Selection with some of its Applications*, Macmillan, London.
Watson, J. D., 1968, *The Double Helix. A Personal Account of the Discovery of the Structure of DNA*, Weidenfeld & Nicholson, London.
Watson, J.D. and Tooze, J., 1981, *The DNA Story. A documentary history of gene cloning*, W.H. Freeman, San Francisco.
Wilson, Edward O., 1975, *Sociobiology. The New Synthesis*, Belknap Press of Harvard University Press, Cambridge (Mass.).
Wilson, Edward O., 1978, *On Human Nature*, Harvard University Press, Cambridge (Mass.).
Wilson, Edward O. and Lumsden, Charles J., 1981, *Genes, Mind and Culture: The Coevolutionary Process*, Harvard University Press, Cambridge (Mass.).
Wilson, Edward O. and Lumsden, Charles J., 1983, *Promethean Fire, Reflections on the Origin of the Mind*, Harvard University Press, Cambridge (Mass.).
Wright, Robert, 1994, *The Moral Animal. Evolutionary Psychology and Everyday Life*, Pantheon Books, New York.

# INDEX

GPSR Compliance
The European Union's (EU) General Product Safety Regulation (GPSR) is a set of rules that requires consumer products to be safe and our obligations to ensure this.

If you have any concerns about our products, you can contact us on

ProductSafety@springernature.com

In case Publisher is established outside the EU, the EU authorized representative is:

Springer Nature Customer Service Center GmbH
Europaplatz 3
69115 Heidelberg, Germany

www.ingramcontent.com/pod-product-compliance
Ingram Content Group UK Ltd.
Pitfield, Milton Keynes, MK11 3LW, UK
UKHW020135160926
475607UK00003B/61
* 9 7 8 0 3 0 6 4 6 6 1 6 8 *